Synthesis Lectures on Information Concepts, Retrieval, and Services

Series Editor

Gary Marchionini, School of Information and Library Science, The University of North Carolina at Chapel Hill, Chapel Hill, USA

This series publishes short books on topics pertaining to information science and applications of technology to information discovery, production, distribution, and management. Potential topics include: data models, indexing theory and algorithms, classification, information architecture, information economics, privacy and identity, scholarly communication, bibliometrics and webometrics, personal information management, human information behavior, digital libraries, archives and preservation, cultural informatics, information retrieval evaluation, data fusion, relevance feedback, recommendation systems, question answering, natural language processing for retrieval, text summarization, multimedia retrieval, multilingual retrieval, and exploratory search.

Weili Guan · Xuemeng Song ·
Dongliang Zhou · Liqiang Nie

Advanced Multimodal Compatibility Modeling and Recommendation

Third Edition

 Springer

Weili Guan
School of Electronics and Information
Engineering
Harbin Institute of Technology
Shenzhen, Guangdong, China

Dongliang Zhou
School of Electronics and Information
Engineering
Harbin Institute of Technology
Shenzhen, Guangdong, China

Xuemeng Song
School of Computer Science and Technology
Shandong University
Qingdao, Shandong, China

Liqiang Nie
Department of Computer Science
and Technology
Harbin Institute of Technology
Shenzhen, Guangdong, China

ISSN 1947-945X   ISSN 1947-9468 (electronic)
Synthesis Lectures on Information Concepts, Retrieval, and Services
ISBN 978-3-031-81047-3   ISBN 978-3-031-81048-0 (eBook)
https://doi.org/10.1007/978-3-031-81048-0

Preface

As society evolves, fashion is not merely a means to clothe the body but an essential facet of human culture, identity, and self-expression. Creating visually harmonious outfits has been a daily routine for people in modern society. However, crafting harmonious ensembles involves more than just following the latest trends; it requires a deep understanding of how different fashion elements—such as color, texture, style, and proportion—interact to form a cohesive and aesthetically pleasing look. For many, achieving this harmony can be daunting, given the vast array of choices available in today's fashion market. The emergence of fashion compatibility modeling represents a significant advancement in assisting individuals with navigating this complexity. By leveraging cutting-edge multimodal processing technologies, fashion compatibility modeling provides a systematic approach to evaluating the compatibility of a set of fashion garments and enables fashion recommendation. This book aims to explore these innovative methods, offering insights into how advanced neural networks, knowledge-driven frameworks, and multimodal integration can revolutionize fashion compatibility assessments and recommendations.

Throughout the book, we address the multifaceted challenges inherent in fashion compatibility modeling. These include integrating the category modality of fashion items, understanding the practical implications of try-on appearance, and the need for fine-grained fashion compatibility modeling. We also explore the challenges of incorporating side information, promoting the model efficiency, as well as the challenges involved in generating compatible fashion items. Each challenge is met with a targeted solution, ranging from category-aware attention networks to sophisticated generative models, all designed to enhance our understanding of fashion compatibility. Our exploration is structured across six chapters, each dedicated to a specific aspect of fashion compatibility modeling. From the foundational principles of multimodal compatibility modeling to the advanced techniques of generating multiple compatible items, the book provides a comprehensive overview of the latest developments in this field. By consolidating these methods, we aim to advance both the academic study and practical application of fashion compatibility modeling, ultimately empowering individuals to make more informed and stylish clothing choices.

We invite you to embark on this journey through the intersection of fashion, technology, and creativity. Whether you are a researcher, a practitioner, or simply a fashion enthusiast, we hope this book will provide valuable insights and inspire new perspectives on the ever-evolving world of fashion.

Shenzhen, China	Weili Guan
Qingdao, China	Xuemeng Song
Shenzhen, China	Dongliang Zhou
Shenzhen, China	Liqiang Nie
October 2024	

Acknowledgements

This book would not have been completed, or at least not in its current form, without the invaluable support of many colleagues, especially those from the iLearn Centers at the Harbin Institute of Technology, Shenzhen, and Shandong University. We would like to take this opportunity to express our deepest gratitude for their contributions to this extensive and time-consuming project.

First, we extend our sincere thanks to several colleagues who made significant contributions to specific chapters of this book: Dr. Peiguang Jing from Tianjin University, and Dr. Xue Dong and Dr. Na Zheng from Shandong University. Their participation in technical discussions, along with their constructive feedback and insightful comments, greatly benefited the development of this book.

Second, we are deeply grateful to the anonymous reviewers, who read the manuscript with great care and provided many thoughtful and constructive suggestions. Their input significantly enhanced the quality of the book.

Third, we sincerely thank Springer Nature Publisher, particularly Dr. Gary Marchionini, the editor, Susanne Filler, and Jayarani Premkumar, the production editors, for their invaluable suggestions and support. Their efforts made the publishing process both smooth and enjoyable.

Finally, we reserve our heartfelt thanks for our beloved families, whose selfless consideration, endless love, and unwavering support made this endeavor possible.

Contents

Introduction

1

1.1 Background

The fashion industry is a dynamic and ever-evolving sector, continuously influenced by cultural transformations, technological advancements, and shifting consumer preferences. Globally, it remains a significant economic force. According to the McKinsey State of Fashion 2024 report,[1] the fashion industry is expected to continually grow despite challenges such as geopolitical instability and inflation. This resilience is driven by sustained consumer demand and the industry's capacity for adaptation, including innovations in sustainability and digital fashion. In this context, the art of crafting harmonious outfits has become increasingly crucial.

Fashion transcends its basic function of covering the body; it serves as a powerful medium for self-expression, enabling individuals to communicate their identity, mood, and personality. A visually-collocated ensemble can boost confidence, convey specific messages, and leave lasting impressions. Achieving such harmony in clothing selection requires more than just a superficial awareness of trends. It requires a comprehensive understanding of how various fashion elements interact to form a cohesive look. The concept of fashion compatibility is central to the creation of visually appealing outfits. This concept involves an understanding of how factors such as color, texture, style, and proportion can be integrated to produce a unified and aesthetically pleasing appearance. While some individuals may have an innate sense of what combinations are visually effective, many others find it challenging to assemble outfits that achieve balance and harmony. The vast array of choices in the modern fashion market can be overwhelming, even for those with a keen sense of style. To address these challenges, the task of fashion compatibility modeling has emerged, focusing on the development of systematic approaches to assess outfit compatibility, assist-

---

[1] https://www.mckinsey.com/industries/retail/our-insights/state-of-fashion.

W. Guan et al., *Advanced Multimodal Compatibility Modeling and Recommendation*,
Synthesis Lectures on Information Concepts, Retrieval, and Services,
https://doi.org/10.1007/978-3-031-81048-0_1

ing individuals in making more informed and aesthetically sound fashion choices. Fashion compatibility modeling presents inherent complexities due to the need of integrate various modalities—such as visual, textual, and categorical information, of fashion items. As each of these modalities significantly influences the perception of compatibility, existing methods focus on multimodal compatibility modeling.

1.2   Challenges

Although early studies have achieved remarkable progress, they suffer from the following key challenges.

The first prominent challenge lies in fully leveraging category information for outfit recommendations. Existing studies have focused solely on unified representation learning across various modalities, neglecting to effectively harness the category information as guidance to improve the dynamic representation of these items.

The second challenge is associated with the practical try-on appearance. Although numerous multimodal compatibility modeling methods yield promising results, they frequently overlook the practical considerations of fashion compatibility—specifically, how garments look when worn. This involves understanding the spatial arrangement and partial coverage of items according to their function or category. Nevertheless, evaluating compatibility based on actual try-on appearance is inherently challenging, especially in the absence of real try-on images.

Another significant challenge is fine-grained fashion compatibility modeling. In fact, fashion compatibility among items is typically influenced by multiple factors such as color, material, and style. Investigating these factors and fulfilling the fine-grained visual compatibility modeling is crucial for a deep understanding of fashion compatibility. However, we do not have explicit labels for supervising these factors modeling. Therefore, how to fulfill accurate fine-grained compatibility modeling without explicit supervision is challenging.

The incorporation of side information presents yet another critical challenge. While some methods enhance fashion item recommendations through visual features, many overlook additional side information, such as attributes like color and category. These attributes often convey key features not fully captured by an item's visual content alone, including material and brand. Integrating semantic attributes can promote the item characteristic learning and user preference learning, and hence being crucial for making accurate recommendations.

The fifth challenge lies in the model efficiency. Previous efforts focus on improving the recommendation effectiveness but overlook the model efficiency. Typically, modeling heterogeneous fashion entities—including users, outfits, items, and attributes—requires integrating these diverse elements within a unified graph. Traditional graph convolution methods often struggle with the inefficiencies facing with such a graph with diverse entity types and entity relationships. On the other hand, directly using conventional hashing for boosting model efficiency may cause much information loss. Therefore, how to boost the model efficiency while maintaining the original data information as much as possible is another challenge.

The final challenge pertains to the compatible item generation, especially generating multiple compatible items simultaneously for a given item represents a notable challenge. Current methods are generally restricted to generating a single image of a compatible clothing item at a time, often depending on general image-to-image translation frameworks. In contrast, methods based on multi-modal image-to-image translation may encounter difficulties when there is significant spatial misalignment between source and target domains, often requiring additional information, such as attributes, user preferences, or reference masks, to guide the image generation process effectively.

1.3   Our Solutions

In response to the complex challenges of fashion compatibility modeling and recommendation, this book explores innovative approaches that introduce advanced methods to overcome the limitations of existing systems. To enhance the accuracy and efficiency of fashion compatibility assessments, we utilize multimodal data, advanced neural network architectures, and knowledge-driven frameworks. By combining theoretical insights with practical applications, this book offers a comprehensive examination of how cutting-edge technology can transform the way we perceive and interact with fashion. Ultimately, it empowers individuals to make more informed and stylish clothing choices.

To tackle the challenge of category-aware compatibility, we introduce a novel category-aware multimodal attention network, which enhances the compatibility modeling of fashion items by utilizing both category-specific visual and textual descriptions. This methodology improves the precision of fashion recommendations by considering contextual nuances inherent in different fashion categories.

To further advance the contextual understanding of fashion compatibility, we present the TryonCM2 framework, which uniquely combines discrete item interaction modeling with combined try-on appearance modeling. This integration allows for a more holistic view on how fashion items interact with one another, providing deeper insights into contextual fashion dynamics.

To address the challenge of fine-grained fashion compatibility modeling, we propose the CTO-Net approach, which also evaluates fashion compatibility from both discrete collocation and try-on perspectives. In particular, it presents a new disentangled graph learning scheme to investigate fine-grained fashion compatibility.

To enhance fashion item recommendation, we introduce MM-FRec, a multi-modal recommendation scheme that utilizes both visual and attribute data. This approach has been validated through extensive experimentation, demonstrating its effectiveness in improving the relevance and appeal of fashion item recommendations.

For addressing the challenge of computational efficiency, we explore a hashing-based outfit recommendation approach. This technique integrates diverse entity types and their relationships through a bi-directional graph convolution algorithm, significantly reducing computational costs while maintaining high-quality recommendations.

Additionally, to address the challenge of generating compatible fashion items, we present BC-GAN, a generative framework capable of synthesizing multiple compatible fashion items simultaneously. By employing contrastive learning, BC-GAN enhances the compatibility between original and synthesized items, a capability validated through experiments on the newly constructed DiverseOutfits dataset.

By consolidating these innovative methods, this book significantly advances the field of fashion compatibility modeling and recommendation. It provides researchers and practitioners with comprehensive insights into leveraging multimodal data, advanced neural networks, and knowledge-guided frameworks to develop more accurate and efficient fashion compatibility and recommendation systems.

1.4   Book Structure

The remainder of this book is organized into six chapters, each dedicated to a specific aspect of advanced multimodal compatibility modeling and recommendation systems in the fashion domain. Chapter 2 explores category-guided fashion compatibility modeling by introducing a category-aware multimodal attention network and a categorical dynamic graph convolutional network. These techniques are designed to enhance visual-semantic embeddings. Chapter 3 delves into try-on-enhanced fashion compatibility modeling through the TryonCM2 framework, where the combined fashion compatibility is modeled by generating a try-on template based on the given fashion items and analyzing its contextual structure with bidirectional long-short term memory (Bi-LSTM) network. Chapter 4 is dedicated to fine-grained and try-on-enhanced outfit compatibility modeling, utilizing the CTO-Net framework. This chapter introduces a disentangled graph learning approach coupled with integrated distillation learning to comprehensively evaluate compatibility from both collocation and try-on perspectives. Chapter 5 presents the MM-FRec framework for attribute-enhanced fashion item recommendation, which jointly models the attribute-enhanced latent preference and visual preference of a user based on GCN. In particular, a multitask learning-based relation-aware propagation method is designed to distinguish importances of neighbor connections with different relation types. Chapter 6 focuses on hashing-based efficient outfit recommendation. It describes a method grounded in heterogeneous graph learning, where a bi-directional graph convolution algorithm is introduced to optimize computational efficiency. To address information loss, dual similarity-preserving regularizations are proposed. The final chapter discusses a collocated clothing synthesis method using the BC-GAN framework. This approach can synthesize multiple collocated fashion items through contrastive learning techniques, with empirical validation performed on the DiverseOutfits dataset. Each chapter builds progressively on the concepts introduced in the previous ones, offering a comprehensive exploration of the latest methods in multimodal compatibility modeling and recommendation within the fashion sector.

Category-Guided Fashion Compatibility Modeling

2

2.1 Introduction

In this chapter, we explore the automatic modeling of fashion compatibility based on fashion categories. The recent surge in the online fashion industry has spotlighted the importance of clothing matching, which holds significant commercial value and high research potential. Although fashion clothing plays an important role in people's daily lives, not everyone is an expert in matching clothes with a keen sense of aesthetics and harmony, which makes composing suitable outfits a tedious daily routine. Moreover, the online fashion industry also needs more intelligent solutions to accommodate the growing demands of users. This challenge thus motivates us to explore effective strategies to automatically estimate the compatibility of fashion items. In fact, fashion items are randomly collected by online fashion communities, such as Polyvore[1] and Chictopia,[2] and usually contain multiple modality information, e.g., visual images, textual descriptions, and annotated categories. Even though each fashion outfit is expressed by various modalities, it is still difficult to determine whether the items are compatible or not if only single modalities are considered. Fortunately, when taking textual descriptions and annotated categories into account, we can explicitly distinguish that a hooded sweatshirt is more compatible with skinny jeans, but a blouse with short sleeves is not suitable for a skater mini skirt, as shown in Fig. 2.1. Inspired by the positive effect of multimodal interaction, many pioneering studies [23, 37] have attempted to utilize the multimodal information of fashion items to better model fashion compatibility.

Despite the acknowledged contribution of current studies, they still have the following limitations: (i) some works either aimed to develop straightforward strategies to explore the complementarity of different modalities or neglected the usability of individual modalities.

---

[1] https://polyvore.ch.
[2] http://www.chictopia.com.

© The Author(s), under exclusive license to Springer Nature Switzerland AG 2025
W. Guan et al., *Advanced Multimodal Compatibility Modeling and Recommendation*,
Synthesis Lectures on Information Concepts, Retrieval, and Services,
https://doi.org/10.1007/978-3-031-81048-0_2

The Type of Outfit Composition	Visual Image	Textual Description	Annotated Category
 **Set_1** — Compatible		Chicnova fashion pure color hooded sweatshirt	>Women's Fashion >Tops>Sweatshirts & Hoodies>**Sweatshirts**
		Topshop moto pretty blue skinny jeans	>Women's Fashion >Clothing>**Jeans**
 **Set_2** — Incompatible		Washed denim blouse with short sleeves	>Women's Fashion >Clothing>Tops >**Blouses**
		Red daisy print skater skirt	Women's Fashion >Clothing>Skirts >**Mini Skirts**

Fig. 2.1 Examples of compatible and incompatible outfits with multimodal metadata (e.g., image, text, and category)

For example, Li et al. [23] and Zhan et al. [52] both adopted simple pooling approaches to obtain multimodal representations of fashion items. Tan et al. [41] and Lin et al. [26] designed condition weight branches to learn the subspace attentions but could not ensure that the learned multimodal representations are sufficiently rich; (ii) many studies [10, 45] have focused only on the unified representation learning of various modalities information while lacking further employment of category information. For example, Vasileva et al. [42] considered exploiting different types of subspaces to learn type-aware embeddings of fashion items but ignored the guidance of categories in assisting the dynamic representations of fashion items; and (iii) similar to aesthetics, sentiment, and memorability, fashion compatibility involves complex and abstract factors, which are challenging to discover in the original space. For example, Chen et al. [3] considered modeling fashion compatibility in Euclidean space, which ignores exploring latent elements behind fashion matching.

In this work, we continue fashion compatibility modeling from the perspective of multimodal representation learning but consider the following problems. (1) Obviously, the visual modality has a dominant effect on fashion compatibility modeling, because it can provide comprehensive and intuitive content descriptions from low-level visual features (e.g., colour, texture, and shape) to high-level visual semantics (e.g., style and pattern). Therefore, a natural solution is to integrate different visual cues based on the hierarchical architecture of convolutional neural networks (CNNs), in which higher semantic cues can be achieved with deeper layers. (2) The category information as high-level attributes can not only be used to make representations of individual modalities category-aware but also provide a bridge to guide the correlation and fusion of multimodal information. Different from existing studies [6, 22], we are committed to adaptively capturing the correlations and representations of categories, which are always varied for different fashion items. Accordingly, how to effectively and dynamically exploit clothing category information is also a primary challenge. (3) In general, multimodal information is complementary to each other. Thus, generating more robust and comprehensive multimodal representations in a mutually beneficial manner still remains quite challenging. Furthermore, it is uncertain which factors contribute more to the underlying relationships between fashion items, so how to adaptively capture significant factors to benefit fashion compatibility modeling is another tough challenge.

To address the above issues, we introduce a fashion compatibility modeling scheme with a category-aware multimodal attention network (FCM-CMAN), which involves deep multimodal feature extraction, categorical dynamic graph convolutional, category-aware contextual attention, and multi-layered multi-head attention networks. Figure 2.2 shows an illustration of our proposed method for fashion compatibility modeling. First, in the deep multimodal feature extraction network, we exploit three separate subnetworks, i.e., CNN [15, 36, 51], TextCNN [35], and GloVe [32], to obtain the initialized visual, textual, and category representations, respectively. In particular, to better explore the power of visual content cues, we feed the outputs of multiple intermediate convolutional layers of the CNN into a global average pooling layer for multi-layered visual representations. Because of the varied category

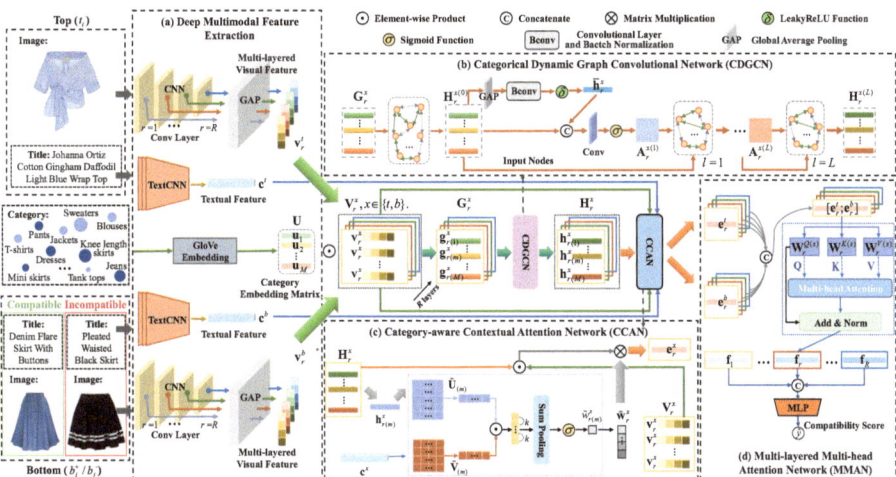

Fig. 2.2 Illustration of the proposed FCM-CMAN framework, which includes the deep multimodal feature extraction, the categorical dynamic graph network, the category-aware contextual attention network, and the multi-layer multi-head attention network. To fully utilize multimodal information, our work mainly proposes a category-aware contextual attention mechanism, which generates visual-semantic embeddings of fashion items to predict fashion compatibility

correlation patterns for different fashion items, we design a categorical dynamic graph convolutional network to adaptively learn multiple content-aware category representations to characterize fashion items from different aspects. Considering the irrelevant information commonly presented in the textual modality, a category-aware contextual attention network intrinsically refines the original contextual representations to better learn context-aware attention weights. Finally, both content-aware representations and context-aware attention weights are integrated by a multimodal factorized bilinear pooling strategy to generate visual-semantic embeddings, which are further improved by a multi-layered multi-head attention network to adequately capture more informative elements. Extensive experiments conducted on two publicly available fashion datasets demonstrate the effectiveness of our proposed method in fashion compatibility modeling.

The main contributions are summarized as follows:

1. We propose a novel category-aware multimodal attention network for fashion compati-bility modeling tasks, in which multi-layered visual and textual descriptions are enforced to be category-aware to better benefit the multimodal representations of fashion items.
2. A categorical dynamic graph convolutional network is fully leveraged to guide the gener-ation of content-aware category representations and contextual-aware attention weights, which are further integrated by a multimodal factorized bilinear pooling operation and a multi-head self-attention mechanism to ensure the effectiveness of visual-semantic embeddings.
3. Extensive experiments conducted on the FashionVC and Expfashion datasets demon-strate that our proposed FCM-CMAN achieves superior performance in comparison with the state-of-the-art methods.

The rest of this chapter is organized as follows. Section 2.2 briefly reviews the related work. The details of our proposed FCM-CMAN are presented in Sect. 2.3. The experimental results and analyses are reported in Sect. 2.4, followed by the conclusion and future work in Sect. 2.5.

2.2   Related Work

2.2.1   Fashion Compatibility Modeling

The flourishing development of the online fashion industry has attracted increasing attention in the field of fashion analysis, such as compatibility modeling [17, 18], fashion retrieval [9, 21], and fashion recommendation [4, 5, 28]. Existing fashion compatibility modeling studies can be approximately divided into the following five groups: (i) the pairwise-based methods [13, 29, 39, 40] generally focus on modeling the fashion compatibility of pairwise items while apparently ignoring the context relationship of fashion items. For example, Han et al.

[13] proposed a prototype-guided interpretable compatibility modeling scheme, which integrates latent prototype learning between fashion items and a Bayesian Personalized Ranking framework; (ii) the sequence-based methods [8, 14, 19] consider fashion outfit as a sequence of items with a fixed predefined order. For example, Han et al. [14] employed a bidirectional long short-term memory (LSTM) network to sequentially model the compatibility relationships among fashion items. Dong et al. [8] designed a try-on-enhanced fashion compatibility modeling framework, which can capture the contextual structure of the try-on appearance by a bidirectional LSTM network. However, these methods often depend on the predefined order of fashion items in an outfit; (iii) the knowledge distillation-based methods [12, 38] utilize fashion domain knowledge rules to guide neural compatibility modeling in a teacher-student network scheme. The limitation of knowledge-guided neural networks is that they may ignore the complex compatibility patterns under inherent knowledge; (iv) the graph-based methods [6, 7, 22, 27] have been developed for fashion compatibility modeling based on graph neural networks. For example, Li et al. [22] proposed a hierarchical graph network to model the relationships among users, items, and outfits. Liu et al. [27] introduced a neural graph filtering framework, which treats fashion items in an outfit as graph nodes to explore the implicit inter-garment relationships; and (v) the category-aware-based methods [26, 42, 49] explores complex similarity relationships by learning category-aware embeddings. For example, Vasileva et al. [42] designed a fashion compatibility modeling scheme to learn the type-respecting embedding for each category-based subspace, and jointly model the similarity and compatibility of items. Besides, Lin et al. [26] and Yang et al. [49] both presented a category-based subspace attention network to capture multiple dimensions of similarities, which integrates category subspace attention mechanism into outfit recommendation task.

Differently, we regard the clothing categories as high-level attributes to refine multimodal representations of fashion items and sufficiently explore a dynamic graph convolutional network to generate content-aware category representations and context-aware attention weights.

2.2.2   Multimodal Representation Learning

Multimodal representation learning [1, 33] is to build models that can process and relate information from multiple modalities. Due to the diversity of data from different sources, multimodal representation aims to reduce the heterogeneity gap between modalities and maintain the integrity of the specific semantics in each modality.

Generally, multimodal representation learning can be roughly divided into two groups, i.e., joint representation learning [16, 46, 48] and coordinated representation learning [2, 31, 44]. The former projects multimodal data into a commonly shared representation space, while the latter processes multimodalities separately into the corresponding representation spaces and reconciles them through additional constraints. For example, Huang et al. [16] provided a joint embedding model based on the triplet network, which integrates multiple types of social relations into multimodal representation learning. Wang et al. [44] proposed a compact hash coding method for multimodal representations, which exploits the intra-modality and inter-modality correlation and incorporates complexities of different modalities. Recently, many compatibility modeling studies have been based on multimodal representation learning, such as SCE-Net [41], MM-OCM [10], and TransNFCM [47]. The multimodal fusion strategies in [10, 47] are relatively simple and the learned visual semantic embedding in [41] considers the inter-modal consistency but neglects the complementarity between modalities. In contrast, we aim to improve the usability of individual modalities based on a contextual attention mechanism and propose a categorical dynamic graph convolutional network to better guide the multimodal representation learning.

2.3   The Proposed Approach

In general, fashion items involve different modalities, e.g., visual images, textual descriptions, and annotated categories, which can be used to characterize their rich information from different aspects. In this work, we mainly explore how to effectively integrate multi-modalities for fashion compatibility modeling. Assume that we have a set of top items $\mathcal{T} = \{t_1, t_2, \ldots, t_{N_t}\}$ and bottom items $\mathcal{B} = \{b_1, b_2, \ldots, b_{N_b}\}$, collected from M fine-grained categories, where N_t and N_b denote the total number of tops and bottoms, respectively. We model each fashion item $t_i \in \mathcal{T}$ ($b_j \in \mathcal{B}$) as a triplet-tuple (v_i^t, c_i^t, u_i^t) $\left((v_j^b, c_j^b, u_j^b) \right)$, where v_i^t (v_j^b), c_i^t (c_j^b), and u_i^t (u_j^b) represent visual image, textual description, and category label, respectively. Based on these items, we can gather a set of positive top-bottom pairs $\mathcal{S} = \{(t_{i_1}, b_{j_1}^+), (t_{i_2}, b_{j_2}^+), \ldots, (t_{i_N}, b_{j_N}^+)\}$ from the expert annotations in downstream tasks, where N is the total number of positive pairs. Accordingly, each top item t_i has a set of positive bottom items $\mathcal{B}_i^+ = \{b_j^+ \in \mathcal{B} | (t_i, b_j^+) \in \mathcal{S}\}$. In addition, we let y_{ij} represent the fashion compatibility label of (t_i, b_j), which can indicate whether the outfit is compatible or not. Specifically, $y_{ij} = 1$ denotes that the top-bottom pair $(t_i, b_j) \in \mathcal{S}$ is compatible, and $y_{ij} = 0$ otherwise.

In this research, we focus on devising a multimodal fashion compatibility modeling scheme $\mathcal{M}$, which is capable of predicting the fashion compatibility score for a given fashion outfit based on multimodal information as follows:

$$\hat{y}_{ij} = \mathcal{M}\left(v_i^t, v_j^b, c_i^t, c_j^b, u_i^t, u_j^b | \Theta \right), \tag{2.1}$$

where $\hat{y}_{ij}$ denotes the predicted compatibility score of a fashion outfit (t_i, b_j). Θ is the set of learnable parameters. Table 2.1 summarizes the main notations used in this work.

Table 2.1 Summary of the main notations

Notation	Explanation
t_i, b_j	The i-th top item and the j-th bottom item
S	The set of positive top-bottom pairs
y_{ij}	The fashion compatibility label of top-bottom pair (t_i, b_j)
$\mathbf{v}_r^x$	The visual feature of the r-th convolutional layer for fashion item x
$\mathbf{c}^x$	The textual feature of fashion item x
$\mathbf{u}_m$	The m-th category embedding
$\mathbf{V}_r^x$	The visual reformulated matrix of $\mathbf{v}_r^x$
$\mathbf{H}_r^{x(l)}$	The content-aware category representation matrix in the l-th graph convolutional layer of $\mathbf{V}_r^x$
$\mathbf{A}_r^{x(l)}$	The adjacency matrix of $\mathbf{H}_r^{x(l)}$
$\tilde{\mathbf{w}}_r^x$	The context-aware attention weights of $\mathbf{v}_r^x$
$\mathbf{e}_r^x$	The visual-semantic embedding of the r-th convolutional layer for fashion item x

2.3.1   Deep Multimodal Feature Extraction

Three separate feature extraction networks are exploited to characterize the following visual, textual, and categorical information, respectively. To improve the readability of our descriptions, we use $x \in \{t, b\}$ to represent a random item, where t and b denote the top and bottom items, respectively.

Multi-layered Visual Feature. To extract more comprehensive visual features of fashion items, we utilize the hierarchy structure of CNNs, which incorporates various convolutional layers to ensure distinct levels of image content descriptions. One typical characteristic of CNNs is that shallow layers can capture more low-level visual cues, while deep layers can deliver more high-level semantics. Considering the influence of different levels visual features on fashion compatibility, we construct the multi-layered visual feature extraction network to characterize fashion items from different convolutional layers, which is similar to previous works [37, 45]. Specifically, we adopt the Global Average Pooling operation (GAP) [24] for the outputs of each convolutional layer, aiming to aggregate the spatial information and map to the features of the specified dimension in a latent space. Formally, given the visual image of the fashion item x, i.e., v^x, we can obtain the corresponding multi-layered visual feature as follows,

$$\mathbf{v}_r^x = f_{\text{GAP}}(\text{Conv}_r(v^x)), r \in \{1, 2, \dots, R\}, \tag{2.2}$$

where $\mathbf{v}_r^x \in \mathbb{R}^{d_v}$ denotes the visual feature of the r-th convolutional layer, d_v is the feature dimension, and R is the number of convolutional layers. $f_{\text{GAP}}(\cdot)$ is the GAP layer and Conv_r is the r-th convolutional layer of CNNs.

Textual Feature. To extract the textual features of fashion items, we utilize the TextCNN [11, 35] model, which has been proven to be robust and effective for numerous natural language processing tasks. For the textual description of each fashion item, we first transform the sentence into a word matrix using the pre-trained word2vec embeddings [30]. We then adopt a single-channel CNN consisting of a convolutional layer and a max pooling layer to extract relevant textual feature and select semantic information. Formally, given the textual description of the fashion item x, i.e., c^x, we can obtain the corresponding textual features as follows,

$$\mathbf{c}^x = \text{TextCNN}(c^x), \tag{2.3}$$

where $\mathbf{c}^x \in \mathbb{R}^{d_c}$ denotes the textual feature of x, and d_c is the feature dimension.

Category Embedding. The category information of fashion items can be regarded as a high-level attribute, which play crucial roles in determining the compatibility of fashion outfits. To make full use of this kind of information, we aim to learn the category embeddings in a dynamic manner to better guide multimodal representation learning. Specifically, we adopt the GloVe word characterization tool [32] to generate initialized category embeddings of M fine-grained categories. Formally, given the m-th category label u_m, we can obtain the category embedding as follows,

$$\mathbf{u}_m = \text{GloVe}(u_m), m \in \{1, 2, \ldots, M\}, \tag{2.4}$$

where $\mathbf{u}_m \in \mathbb{R}^{d_u}$ denotes the embedding of the m-th category, and d_u is the embedding dimension.

2.3.2 Categorical Dynamic Graph Convolutional Network

Since category correlations are always dynamic and varied for various fashion items, we present a categorical dynamic graph convolutional network (CDGCN) to adaptively capture the semantic correlations between categories. To learn these dynamic category correlations, we first define a dynamic graph $G_d = (\mathbf{U}, \mathbf{A})$ with node feature matrix $\mathbf{U} \in \mathbb{R}^{M \times d_u}$ and adjacency matrix $\mathbf{A} \in \mathbb{R}^{M \times M}$. In our case, we regard each node as one of the predefined category labels, and the adjacency matrix reflects the interactive relationships among the nodes.

Initialized Category Representation. To enhance the content-aware representation learning of visual feature for category embedding, we first use the category embedding matrix $\mathbf{U} = [\mathbf{u}_1, \ldots, \mathbf{u}_m, \ldots, \mathbf{u}_M]^T \in \mathbb{R}^{M \times d_u}$ to perform the Hadamard product operation with the visual reformulated matrix $\mathbf{V}_r^x = [\mathbf{v}_r^x, \mathbf{v}_r^x, \ldots, \mathbf{v}_r^x]^T \in \mathbb{R}^{M \times d_v}$, which is generated by repeating the same visual feature vector M times. Concretely, for the fashion item x, we can produce the content-based category representation matrix $\mathbf{G}_r^x \in \mathbb{R}^{M \times d_g}$ of the r-th convolutional layer as follows,

$$
\begin{aligned}
\mathbf{G}_r^x &= \mathbf{U} \odot \mathbf{V}_r^x \\
&= [\mathbf{g}_{r(1)}^{x(0)}, \ldots, \mathbf{g}_{r(m)}^{x(0)}, \ldots, \mathbf{g}_{r(M)}^{x(0)}]^T,
\end{aligned}
\tag{2.5}
$$

where $\odot$ denotes the element-wise product operation, $\mathbf{g}_{r(m)}^{x(0)} \in \mathbb{R}^{d_g}$ is the m-th content-based category representation vector of $\mathbf{v}_r^x$, and d_g is the dimension. On this condition, we set $d_g = d_u = d_v$.

We then employ the first graph convolutional layer to formulate the node representation matrix $\mathbf{G}_r^x$ in the dynamic graph. Formally, for the fashion item x, we can obtain the initialized category representation matrix $\mathbf{H}_r^{x(0)} \in \mathbb{R}^{M \times d_h}$ of the r-th convolutional layer as follows,

$$
\begin{aligned}
\mathbf{H}_r^{x(0)} &= \delta(\mathbf{A}_r^{x(0)} \mathbf{G}_r^x \mathbf{W}_r^{x(0)}) \\
&= [\mathbf{h}_{r(1)}^{x(0)}, \ldots, \mathbf{h}_{r(m)}^{x(0)}, \ldots, \mathbf{h}_{r(M)}^{x(0)}]^T,
\end{aligned}
\tag{2.6}
$$

where $\mathbf{A}_r^{x(0)} \in \mathbb{R}^{M \times M}$ is the predefined adjacency matrix, $\mathbf{W}_r^{x(0)} \in \mathbb{R}^{d_g \times d_h}$ is the initial learnable transformation matrix, and $\delta(\cdot)$ denotes the LeakyReLU activation function. Moreover, $\mathbf{h}_{r(m)}^{x(0)} \in \mathbb{R}^{d_h}$ is the m-th initialized category representation vector of $\mathbf{v}_r^x$, and d_h is the dimension.

Category Correlations Update. To adaptively update the dynamic correlation matrix, for the fashion item x, we utilize the GAP and the Bconv operation (includes a convolutional layer and a batch normalization layer) to capture the global spatial information as follows,

$$
\bar{\mathbf{h}}_r^x = \sigma[f_{\text{Bconv}}(f_{\text{GAP}}(\mathbf{H}_r^{x(0)}))],
\tag{2.7}
$$

where $\bar{\mathbf{h}}_r^x \in \mathbb{R}^{d_h}$ is the global category representation of $\mathbf{v}_r^x$, $\sigma(\cdot)$ denotes the sigmoid activation function, and $f_{\text{Bconv}}(\cdot)$ stands for the Bconv operation.

To further prevent the loss of information caused by the increasing number of graph convolutional layers, we are inspired by the residual network architecture [15], propagating the initial global representation remotely via a shortcut connection. Specifically, we concatenate the initialized category representation matrix $\mathbf{H}_r^{x(0)}$ with the obtained global category representation $\bar{\mathbf{h}}_r^x$ as follows,

$$
\begin{aligned}
\mathbf{H}_r^{x(l-1)} = \\
\left[(\mathbf{h}_{r(1)}^{x(l-1)}; \bar{\mathbf{h}}_r^x), \ldots, (\mathbf{h}_{r(m)}^{x(l-1)}; \bar{\mathbf{h}}_r^x), \ldots, (\mathbf{h}_{r(M)}^{x(l-1)}; \bar{\mathbf{h}}_r^x)\right]^T,
\end{aligned}
\tag{2.8}
$$

where $\mathbf{H}_r^{x(l-1)} \in \mathbb{R}^{M \times 2d_h}$ is the intermediate category representation matrix of the l-th graph convolutional layer, and $(;)$ denotes the concatenation operation. In addition, $l \in \{1, 2, \ldots, L\}$, where L is the number of graph convolutional layers. Subsequently, we can dynamically update the adjacency matrix $\mathbf{A}_r^{x(l)} \in \mathbb{R}^{M \times M}$ based on the intermediate category representation as follows,

$$\mathbf{A}_r^{x(l)} = \sigma(\mathbf{H}_r^{x(l-1)} \mathbf{W}_{r(A)}^{x(l)}), \tag{2.9}$$

where $\mathbf{W}_{r(A)}^{x(l)} \in \mathbb{R}^{2d_h \times M}$ is the learnable weight matrix of the l-th graph convolutional layer.

Content-aware Category Representation. Ultimately, for the fashion item x, the intermediate category representation and corresponding correlations derived from the former layer are transmitted to the next layer to learn content-aware category representation matrix $\mathbf{H}_r^{x(l)} \in \mathbb{R}^{M \times d_h}$ as follows,

$$\begin{aligned}
\mathbf{H}_r^{x(l)} &= \delta(\mathbf{A}_r^{x(l)} \mathbf{H}_r^{x(l-1)} \mathbf{W}_r^{x(l)}) \\
&= [\mathbf{h}_{r(1)}^{x(l)}, \dots, \mathbf{h}_{r(m)}^{x(l)}, \dots, \mathbf{h}_{r(M)}^{x(l)}]^T,
\end{aligned} \tag{2.10}$$

where $\mathbf{W}_r^{x(l)} \in \mathbb{R}^{d_h \times d_h}$ is the learnable transformation matrix of the l-th graph convolutional layer, and $\mathbf{h}_{r(m)}^{x(l)} \in \mathbb{R}^{d_h}$ is the m-th content-aware category representation vector of $\mathbf{v}_r^x$. To simplify notation, we use $\mathbf{H}_r^x$ to represent the outputs of the final graph convolutional layer $\mathbf{H}_r^{x(L)}$.

2.3.3 Category-Aware Contextual Attention Network

Considering that the textual description of fashion items can provide contextual information for compatibility modeling, we devote to learning more comprehensive multimodal representations by jointly integrating textual features and content-aware category representations to generate different category subspace weights. Due to the influence of noise in textual descriptions, we propose a category-aware contextual attention network (CCAN) to improve the usability of the contextual representation of fashion items. Inspired by the multimodal factorized bilinear pooling (MFBP) operation [50], for the fashion item x, the m-th content-aware category representation $\mathbf{h}_{r(m)}^x$ and textual feature $\mathbf{c}^x$ are unified to obtain the fused result $\tilde{w}_{r(m)}^x \in \mathbb{R}$ as follows,

$$\tilde{w}_{r(m)}^x = \mathbf{h}_{r(m)}^x{}^T \tilde{\mathbf{W}}_{(m)} \mathbf{c}^x, \tag{2.11}$$

where $\tilde{\mathbf{W}}_{(m)} \in \mathbb{R}^{d_h \times d_c}$ is the learnable projection matrix in terms of the m-th category.

In order to enrich the fused result from a low-rank intrinsic subspace, the projection matrix $\tilde{\mathbf{W}}_{(m)}$ can be factorized and reshaped as two low-rank matrices. Then, Eq. (2.11) is reformulated as follows,

$$\tilde{w}_{r(m)}^x = \text{SumPool}(\tilde{\mathbf{U}}_{(m)}^T \mathbf{h}_{r(m)}^x \odot \tilde{\mathbf{V}}_{(m)}^T \mathbf{c}^x, k), \tag{2.12}$$

where $\tilde{\mathbf{U}}_{(m)} \in \mathbb{R}^{d_h \times k}$ and $\tilde{\mathbf{V}}_{(m)} \in \mathbb{R}^{d_c \times k}$ are the reformulated factorized matrices, and k is the factorization rank. Moreover, the function $\mathrm{SumPool}(x, k)$ applies sum pooling within k nonoverlapping windows to x. Accordingly, context-aware attention weights of $\mathbf{v}_r^x$ can be described as $\tilde{\mathbf{w}}_r^x = \left[\tilde{w}_{r(1)}^x, \ldots, \tilde{w}_{r(m)}^x, \ldots, \tilde{w}_{r(M)}^x \right]^T \in \mathbb{R}^M$.

By combining the content-aware category representations with multi-layered visual features and context-aware attention weights, for the fashion item x, a contextual attention mechanism is developed to aggregate visual-semantic embedding from multiple perspectives as follows,

$$\mathbf{e}_r^x = [\mathbf{H}_r^x \odot \mathbf{V}_r^x]^T \tilde{\mathbf{w}}_r^x, \qquad (2.13)$$

where $\mathbf{e}_r^x \in \mathbb{R}^{d_e}$ is the visual-semantic embedding of the r-th convolutional layer, and d_e is the dimension.

2.3.4 Multi-layered Multi-head Attention Network

Enlightened by the self-attention mechanism [43] in adaptively capturing more informative elements, a multi-layered multi-head attention network is designed to discover significant elements that are related to fashion compatibility. While the visual-semantic embeddings of an individual fashion item can be well explored, the collective actions from different fashion items in each outfit also deserve to be further considered. Regarding the joint consideration of the top and bottom items in a fashion outfit, we first concatenate them to constitute the initial outfit representation $\hat{\mathbf{e}}_r = [\mathbf{e}_r^t; \mathbf{e}_r^b] \in \mathbb{R}^{2d_e}$ of the r-th convolutional layer.

Then a multi-head self-attention (MSA) mechanism is employed to further condense the concatenated results. Specifically, the hidden outfit representation $\mathbf{z}_r^{(s)}$ of the r-th convolutional layer for the s-th head is defined as follows,

$$\mathbf{z}_r^{(s)} = \mathrm{softmax} \left[\frac{\left(\hat{\mathbf{e}}_r \mathbf{W}_r^{Q(s)} \right) \left(\hat{\mathbf{e}}_r \mathbf{W}_r^{K(s)} \right)^T}{\sqrt{d_k}} \right] \left(\hat{\mathbf{e}}_r \mathbf{W}_r^{V(s)} \right), \qquad (2.14)$$

where $\mathbf{W}_r^{Q(s)} \in \mathbb{R}^{2d_e \times d_q}$, $\mathbf{W}_r^{K(s)} \in \mathbb{R}^{2d_e \times d_k}$, and $\mathbf{W}_r^{V(s)} \in \mathbb{R}^{2d_e \times d_v}$ are the linear projection matrices corresponding to the query, key, and value, respectively. d_q, d_k, and d_v are the dimensions of different projection matrices, where $d_q = d_k = d_v$. Additionally, $\mathrm{softmax}(\cdot)$ denotes the softmax activation function.

The outputs of multi-heads are concatenated and aggregated by a linear layer, and then use a residual connection and a normalization layer to obtain the outfit representation $\mathbf{f}_r \in \mathbb{R}^{2d_e}$ of the r-th convolutional layer as follows,

$$\begin{cases} \mathrm{MSA}(\hat{\mathbf{e}}_r) = [\mathbf{z}_r^{(1)}; \mathbf{z}_r^{(2)}; \ldots; \mathbf{z}_r^{(s)}]\mathbf{W}_o, s \in \{1, 2, \ldots, S\}, \\ \mathbf{f}_r = \mathrm{LN}(\hat{\mathbf{e}}_r + \mathrm{MSA}(\hat{\mathbf{e}}_r)), \end{cases} \qquad (2.15)$$

where $\mathbf{W}_o \in \mathbb{R}^{2Sd_e \times 2d_e}$ is the learnable weight matrix, S is the total number of heads in the self-attention mechanism, and $\text{LN}(\cdot)$ denotes the layer normalization operation.

Finally, given the fashion outfit (t_i, b_j), we concatenate the outfit representations of the total R convolutional layers to obtain the final multi-layered fusion outfit representation $\tilde{\mathbf{f}}_{(ij)} = [\mathbf{f}_1; \ldots; \mathbf{f}_r; \ldots; \mathbf{f}_R] \in \mathbb{R}^{2Rd_e}$, which is further fed into the Multi-Layer Perceptron (MLP) with two layers for fashion compatibility prediction as follows,

$$\hat{y}_{ij} = \text{MLP}(\tilde{\mathbf{f}}_{(ij)}), \tag{2.16}$$

where $\hat{y}_{ij}$ is the predicted compatibility score of a fashion outfit (t_i, b_j). Notably, the ReLU activation function is adopted in the front layers to enhance nonlinearity, and the Sigmoid function is used in the last layer to project the compatibility scores into the range of $[0, 1]$.

2.3.5 Objective Function

To optimize the FCM-CMAN, we can present a unified formulation, which includes the classification loss $\mathcal{L}_{clf}$ measuring the divergences between predicted results and ground truth labels, the outfit compatibility ranking loss $\mathcal{L}_{ocr}$ encouraging the compatibility scores of positive outfits to be higher than those of negative outfits, the dynamic graph loss $\mathcal{L}_{cdg}$ controlling the categorical dynamic graph convolutional network, and the embedding loss $\mathcal{L}_{emb}$ encoding the normalized visual-semantic embeddings of fashion items in the latent space.

Classification Loss. The binary cross-entropy loss is widely used in various classification tasks, so we adopt it as our classification loss $\mathcal{L}_{clf}$ as follows,

$$\mathcal{L}_{clf} = -[y_{ij}\log(\hat{y}_{ij}) + (1 - y_{ij})\log(1 - \hat{y}_{ij})], \tag{2.17}$$

where $\hat{y}_{ij}$ and y_{ij} denote the predicted result and the ground truth of a fashion outfit (t_i, b_j), respectively.

Outfit Compatibility Ranking Loss. To explore the implicit compatibility preferences between top and bottom items, we reconstruct the loss of the Bayesian personalized ranking (BPR) framework [34] with the triplet loss. Assuming that bottom items from the positive set $\mathcal{B}_i^+$ are more compatible with top item t_i than those non-composed bottom items, we can build a triplet set $\mathcal{D}_S$ as follows,

$$\mathcal{D}_S := \left\{ (t_i, b_j^+, b_j^-) | t_i \in \mathcal{T}, b_j^+ \in \mathcal{B}_i^+ \wedge b_j^- \in \mathcal{B} \setminus \mathcal{B}_i^+ \right\}, \tag{2.18}$$

where the item triplet (t_i, b_j^+, b_j^-) indicates that the score of the positive outfit (t_i, b_j^+) is higher than that of the negative outfit (t_i, b_j^-). Accordingly, the outfit compatibility ranking loss $\mathcal{L}_{ocr}$ can be defined as follows,

$$\mathcal{L}_{ocr} = \frac{1}{|\mathcal{D}_\mathcal{S}|} \sum_{(t_i, b_j^+, b_j^-) \in \mathcal{D}_\mathcal{S}} \max\left\{0, \hat{y}_{ij}^- - \hat{y}_{ij}^+ + \mu\right\}, \tag{2.19}$$

where $\hat{y}_{ij}^+$ and $\hat{y}_{ij}^-$ denote the fashion compatibility score of the positive outfit (t_i, b_j^+) and the negative outfit (t_i, b_j^-), respectively, and $\mu > 0$ is a certain margin.

Dynamic Graph Loss. To better guide content-aware category representation learning, we suppose that the real and predicted binary category vectors are $\mathbf{h}$ and $\hat{\mathbf{h}} \in \mathbb{R}^M$, respectively. Formally, the dynamic graph loss $\mathcal{L}_{cdg}$ based on the multi-label classification is defined as follows,

$$\mathcal{L}_{cdg} = \sum_{m=1}^{M} h_m \log(\sigma(\hat{h}_m)) + (1 - h_m)\log(1 - \sigma(\hat{h}_m)), \tag{2.20}$$

where the m-th elements h_m and $\hat{h}_m$ represent whether the m-th category label appears in a fashion item or not.

Embedding Loss. To encode the normalized multimodal representation in each convolutional layer, we use the embedding loss $\mathcal{L}_{emb}$ to control the visual-semantic embeddings of fashion items as follows,

$$\mathcal{L}_{emb} = \sum_{r=1}^{R} \left\| e_r^x \right\|_2, \tag{2.21}$$

where $\|\cdot\|_2$ denotes the $\mathcal{L}_2$-norm.

By combining Eqs. (2.17), (2.19), (2.20), and (2.21), we formulate the total objective function $\mathcal{L}_{total}$ as follows:

$$\mathcal{L}_{total} = \lambda_1 \mathcal{L}_{clf} + \lambda_2 \mathcal{L}_{ocr} + \lambda_3 \mathcal{L}_{cdg} + \lambda_4 \mathcal{L}_{emb}, \tag{2.22}$$

where $\lambda_1, \lambda_2, \lambda_3$, and λ_4 are the trade-off parameters to balance the contributions of different losses.

The overall optimization procedure of our proposed FCM-CMAN is summarized in Algorithm 2.1.

Algorithm 2.1 Optimization of FCM-CMAN

Input: Training set $\mathcal{D}_S$; hyper-parameters $\lambda_1, \lambda_2, \lambda_3, \lambda_4$;
prior adjacency matrix of categories $\mathbf{A}^{(0)}$.
Output: Outfit compatibility scores $\hat{y}_{ij}^+, \hat{y}_{ij}^-$ and parameters $\mathbf{\Theta}$.
1: Randomly initialize all network parameters $\mathbf{\Theta}$;
2: **repeat**
3: **for** $epoch = 1, \ldots, N$ **do**
4: Draw the item triplet (t_i, b_j^+, b_j^-) from $\mathcal{D}_S$
5: **for** $r = 1, \ldots, R$ **do**
6: Update the initialized category representation matrix $\mathbf{H}_r^{x(0)}$ with Eqs. (2.5) and (2.6);
7: Update the dynamic category adjacency matrix $\mathbf{A}_r^{x(l)}$ with Eqs. (2.7), (2.8) and (2.9);
8: Update the content-aware category representation matrix $\mathbf{H}_r^{x(l)}$ with Eq. (2.10);
9: Update the context-aware attention weights $\tilde{\mathbf{w}}_r^x$ with Eq. (2.12);
10: Update visual-semantic embedding $\mathbf{e}_r^x$ with Eq. (2.13);
11: Update outfit representation $\mathbf{f}_r$ with Eqs. (2.14) and (2.15);
12: **end**
13: Update multi-layered fusion outfit representation $\tilde{\mathbf{f}}_{(ij)}$
with $\tilde{\mathbf{f}}_{(ij)} = [\mathbf{f}_1; \ldots; \mathbf{f}_r; \ldots; \mathbf{f}_R]$, $r \in \{1, 2, \ldots, R\}$;
14: Update predicted scores $\hat{y}_{ij}^+, \hat{y}_{ij}^-$ with Eq. (2.16);
15: Compute loss with Eq. (2.22) and update the parameters $\mathbf{\Theta}$;
16: **end**
17: **until** Convergence.

2.4   Experiments

2.4.1   Experimental Settings

Datasets. We conducted experiments on two publicly released datasets FashionVC [39] and ExpFashion [25], both of which are collected from the online fashion community Polyvore. **FashionVC** consists of $20, 726$ outfits with $14, 871$ tops and $13, 663$ bottoms, while **ExpFashion** is comprised of $200, 745$ outfits with $29, 113$ tops and $20, 902$ bottoms. Each fashion item contains multimodal data, such as visual image, textual description and annotated categories. Notably, following the same dataset settings as in [12], we sampled $20, 000$ positive outfits in ExpFashion. In our experiments, we utilized this multimodal information and particularly selected 35 fine-grained categories assigned to fashion items. For each dataset, we divided the positive top-bottom pair set $\mathcal{S}$ into the training set $\mathcal{S}_{train}$ (80%), validation set $\mathcal{S}_{valid}$ (10%), and test set $\mathcal{S}_{test}$ (10%). Then, we generated the triple set $\mathcal{D}_{\mathcal{S}_{train}}, \mathcal{D}_{\mathcal{S}_{valid}}$, and $\mathcal{D}_{\mathcal{S}_{test}}$ according to Eq. (2.18). For each positive top-bottom pair (t_i, b_j^+), we randomly sampled 3 bottom items b_j^-'s and each b_j^- contributes to an item triplet (t_i, b_j^+, b_j^-), where $b_j^- \notin \mathcal{B}_i^+$. To better evaluate the estimated accuracy, we adopted the area under the ROC curve (AUC) [53] as the evaluation metric. In addition, to verify the practicality of the proposed model, we extended experiments in the context of complementary fashion item retrieval task. Given a query item, this task utilizes the trained model to select the compatible complementary item from a candidate set.

Implementation Details. For the visual modality, we utilized the pre-trained ResNet-50 [15] as our backbone network and selected the outputs of 4 different convolutional layers i.e., conv2_x, conv3_x, conv4_x, and conv5_x to obtain the 300D multi-layered visual features according to Eq. (2). For the textual modality, we first adopted the pre-trained word2vec tool to obtain the 300D word vectors for each textual description, and fed the concatenation of them into the TextCNN equipped with four convolution kernels in different sizes [2, 3, 4, 5], where each kernel with 100 feature maps and the ReLU activation function. Finally, we extracted the 400D textual feature for each fashion item. In addition, we used the GloVe word characterization tool to produce the 300D category embedding for each category.

In the categorical dynamic graph convolutional network, we adopted a two-layer graph structure to learn the content-aware category representations. The factorization rank of the category-aware contextual attention network is set as $k = 50$ on FashionVC, and $k = 250$ on ExpFashion. Additionally, the number of heads in the multi-layered multi-head attention network is set as $S = 4$.

For the optimization of our proposed FCM-CMAN, adaptive moment estimation (Adam) [20] was used as the optimizer. We fine-tuned the hyper-parameters based on the training set and validation set for 60 epochs, and reported the performance on the test set. FCM-CMAN achieves the optimal performance with an initial learning rate of $5e^{-5}$, which decays by a factor of 0.1 every 15 epochs. And the hyper-parameters λ_1, λ_2, λ_3, and λ_4 in the total loss function are set to 1, 0.6, $1e^{-2}$, and $5e^{-3}$, respectively. In addition, we set the margin value to $\mu = 0.5$ by default for the outfit compatibility ranking loss. All experiments are implemented based on PyTorch and conducted on a single NVIDIA GeForce RTX 3090 GPU.

2.4.2 Performance Evaluation

Convergence Analysis. In this part, we investigated the convergence of our proposed model in terms of the loss function and the evaluation metric AUC. Figure 2.3 visualizes the convergence curves of the loss function and AUC of the two datasets. From the figures, we can observe that the loss curves (blue solid and dashed lines) drop sharply as the number of epochs increases at the beginning stage and then tend to stabilize at the later stage, while the AUC curves (yellow solid and dashed lines) first rise sharply and then tend to become stable, which indicates the good convergence and stability of our proposed model during the training stage.

Comparison with State-of-the-Arts. To illustrate the effectiveness of our proposed FCM-CMAN, we compared it with the following state-of-the-art methods as baselines, including pair-based methods, sequence-based methods, knowledge distillation-based methods, graph-based methods, and category-aware-based methods.

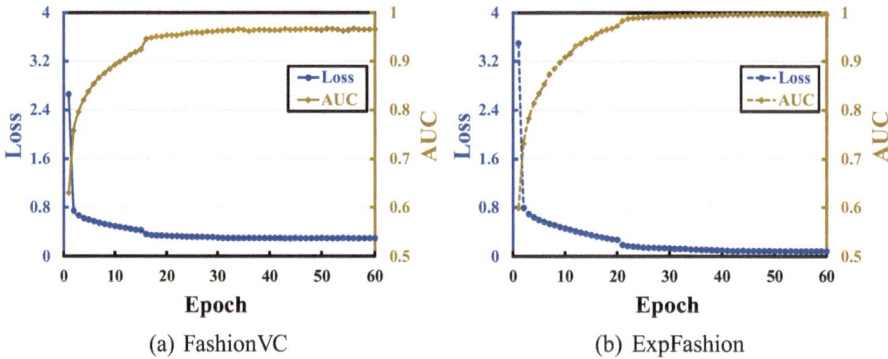

Fig. 2.3 Convergence curves of the loss function and AUC on **a** FashionVC and **b** ExpFashion. The horizontal axis is the number of epochs, and the left and right vertical axes show the values of the loss function and AUC, respectively

- **IBR** [29]: This image-based recommendation method models the relationships between objects in a linear latent style space simply with visual information.
- **ExIBR**: This method is an extension of the IBR introduced by [38] as one baseline, which measures the compatibility of fashion items in latent style spaces with both visual and textual information.
- **BPR-DAE** [39]: By incorporating a dual autoencoder network, the content-based neural method jointly models the implicit preference among fashion items and the coherent relation between multiple modalities.
- **Bi-LSTM** [14]: Regarding an outfit as a sequence, the bidirectional LSTM method explores the sequential relationships among items to learn fashion compatibility. In this research, we adapted Bi-LSTM to deal with outfits that consist of simply a top item and a bottom item.
- **AKD-DBPR-p** [38]: This method is based on attentive knowledge distillation, which uses a teacher-student network to encode knowledge and assigns rule confidences with the attention mechanism. "-p" denotes that the final student network is utilized to estimate the compatibility of fashion items.
- **PKD-DBPR-p** [12]: Based on probabilistic knowledge distillation, this method utilizes a teacher-student network to encode vast knowledge rules in a probabilistic manner. "-p" has the same meaning as above.
- **Context-aware** [6]: By treating fashion compatibility modeling as an edge prediction problem, this method constructs a graph auto-encoder framework to exploit the underlying relational information and learn the contextual embeddings of fashion items.
- **NGNN** [7]: To enhance the representations of fashion items and the modeling of their relations, this method utilizes a node-wise graph neural network and attention mechanism to infer outfit compatibility from multiple modalities, where nodes represent categories and edges denote interactions among nodes.

Table 2.2 Performance comparison of state-of-the-art methods in terms of AUC on FashionVC and ExpFashion

Method	Feature	FashionVC	ExpFashion
IBR	Visual	0.6075	0.6611
ExIBR	Visual, Textual	0.7033	0.7394
BPR-DAE	Visual, Textual	0.7616	0.7912
Bi-LSTM	Visual, Textual	0.6629	0.6717
AKD-DBPR-p	Visual, Textual	0.7843	0.8727
PKD-DBPR-p	Visual, Textual	0.7881	0.8819
Context-aware	Visual	0.7642	0.8397
NGNN	Multi-cues	0.7728	0.8648
SCE-Net	Multi-cues	0.7904	0.8896
MCAN	Visual, Category	0.7943	0.8965
LPAE	Visual	0.7836	0.8723
FCM-CMAN	**Multi-cues**	**0.8380**	**0.9115**

- **SCE-Net** [41]: This method jointly learns multiple similarity condition-aware embeddings from different semantic subspaces, and integrates multimodal features to generate the condition weights. In particular, we incorporated category information as the input of the conditional weight branch.
- **MCAN** [49]: This method designs a mixed category attention network to learn fashion compatibility among multiple tuples, involving fashion items and category pairs. In this research, we only used the fine-grained category as category information.
- **LPAE** [28]: To model the high-order interactions among fashion items, this method exploits a self-attention aggregation network, and then the attentive pooling operation is used to aggregate item embeddings into a single compact outfit representation.

In our experiments, we used the same training, validation, and test splits for all comparison methods. Table 2.2 shows the performance comparison of the state-of-the-art methods on FashionVC and ExpFashion. To facilitate comparison, the utilization of different modalities in all baseline methods is illustrated, where "Multi-cues" indicates that visual, textual and category information are used. From the table, we make the following observations:

- FCM-CMAN outperforms all the comparison methods in terms of AUC on the two datasets, demonstrating its superiority in fashion compatibility modeling.
- Three pair-based methods, i.e., IBR, ExIBP, and BPR-DAE, obtain unsatisfactory performance, indicating that the fashion style spaces and the content-based autoencoder are still insufficient to model the complicated compatibility of fashion items, despite multimodal information exploited in ExIBP and BPR-DAE. Additionally, the

sequence-based method, i.e., Bi-LSTM, also gets worse performance than other base-lines. A possible reason is that the sequence representation lacking auxiliary information may cause cumulative error propagation.

- Two knowledge distillation-based methods, i.e., AKD-DBPR-p and PKD-DBPR-p, sur-pass the pair-based and sequence-based methods, indicating the benefit of knowledge distillation in the context of compatibility modeling. However, some knowledge rules learned are fuzzy and not applicable to guide all cases.

- Two graph-based methods, i.e., Context-aware and NGNN, achieve relatively promising results. NGNN outperforms Context-aware in the graph-based methods, which reflects the advantage of employing multimodal embedding in a category space. Notably, the dynamic graph in our FCM-CMAN can better capture the semantic correlations between categories.

- The category-aware-based method, i.e., SCE-Net and MCAN show the superiority over all baselines, which indicates category-aware embedding learning can comprehensively characterize fashion items. Despite LPAE employs a self-attention aggregation network to model fashion outfit without category guidance, which verifies the effectiveness of the attention mechanism. It is worth noting that our proposed method can achieve better performance by integrating the category-aware representation and contextual attention mechanism.

Parameter Sensitivity Analysis. In this part, we first investigated the influences of the hyper-parameters λ_1, λ_2, λ_3, and λ_4 in the loss function, which are used to control the classification, outfit compatibility ranking, dynamic graph, and embedding losses, respec-tively. In our experiments, λ_1 and λ_2 are selected through a grid search in a heuristic manner with the same interval, while λ_3 and λ_4 change with a 10-time interval. Figure 2.4 shows the performance with various values of the hyper-parameters λ_1, λ_2, λ_3, and λ_4 in terms of AUC on two datasets. Each curve represents the performance when varying the values of one hyper-parameter while the other hyper-parameters remain fixed. From the figures, we find that the best performance is achieved on two datasets when λ_1, λ_2, λ_3, and λ_4 are set to 1, 0.6, $1e^{-2}$, and $5e^{-3}$, respectively. Nevertheless, λ_2 exhibits strong fluctuations in contrast to others, illustrating that our proposed FCM-CMAN is more sensitive to outfit

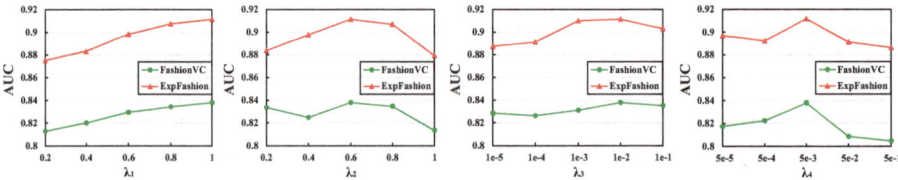

Fig. 2.4 Performance with various values of the hyper-parameters λ_1, λ_2, λ_3, and λ_4 in terms of AUC on two datasets. Each curve indicates that one of these parameters has changed, while other parameters are fixed at the optimal solution

Table 2.3 Performance comparison with different numbers of dynamic graph layers in the CDGCN module in terms of AUC on FashionVC and ExpFashion

# Layer	FashionVC	ExpFashion
1	0.7254	0.8847
2	**0.8380**	**0.9115**
3	0.8152	0.8468
4	0.8293	0.8638

compatibility ranking loss. Meanwhile, we observed that the performance is far worse than the final result when the values of λ_1 and λ_3 are small, indicating the necessity of classification and dynamic graph losses.

To investigate the effectiveness of the involved categorical dynamic graph convolutional network, we considered changing the number of graph layers ranging from 1 to 4. Table 2.3 shows the performance comparison with different numbers of dynamic graph layers in the CDGCN module. As shown in the table, the best performance is obtained when $L = 2$. The performance drops sharply as the number of layers increases, which indicates that the shallow layers of the dynamic graph are sufficient to learn the semantic correlations and the increasing layers can lead to overfitting.

To investigate the effectiveness of the involved category-aware contextual attention network, we considered changing the numbers of factorization rank (k) ranging from 50 to 950. Figure 2.5 shows the performance comparison with different numbers of factorization rank in the CCAN module. From the figure, we find that the best performance is obtained when $k = 50$ on FashionVC and $k = 250$ on ExpFashion, which indicates that the optimal value of factorization rank on the two datasets is obviously different. Meanwhile, the performance gets worse as the number of k increases. One possible reason is that the high-dimensional factorization rank can easily lead to serious overfitting.

Additionally, we further investigated the influences of the number of heads in the multi-layered multi-head attention network by ranging the number from 1 to 6. Table 2.4 shows the performance comparison with different numbers of heads in the MMAN module. From the table, we can find that the best performance is obtained when $S = 4$. As the number of attention heads increases and $S > 4$, the model performance gets worse due to parameter redundancy.

Ablation Analysis. To validate the contribution of each component in our proposed model, we compared the performance of the different schemes by removing the relevant components as follows:

- **noCDGCN**: We eliminated the influence of the categorical dynamic graph convolutional network by exploiting the traditional graph convolutional network to learn the content-aware category representations.

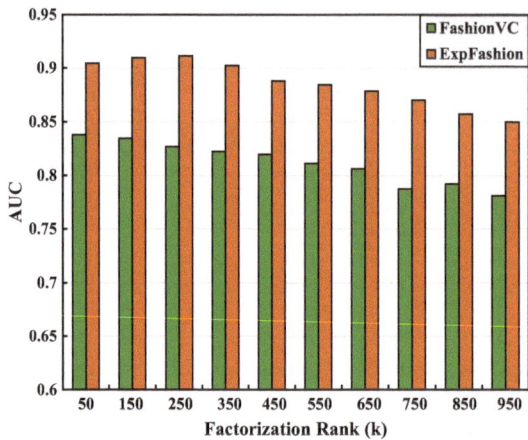

Fig. 2.5 Performance comparison with different numbers of factorization rank (k) in the CCAN module in terms of AUC on FashionVC and ExpFashion

Table 2.4 Performance comparison with different numbers of heads in the MMAN module in terms of AUC on FashionVC and ExpFashion

# Head	FashionVC	ExpFashion
1	0.7242	0.8446
2	0.7492	0.8289
3	0.8232	0.8788
4	**0.8380**	**0.9115**
5	0.7853	0.8524
6	0.7324	0.8375

- **noCCAN**: We eliminated the influence of the category-aware contextual attention network by randomly generating context-aware attention weights.
- **noMMAN**: We eliminated the influence of the multi-layered multi-head attention network by directly concatenating the multi-layered visual-semantic embeddings of fashion items for compatibility prediction.

Table 2.5 shows the performance comparison of different ablation schemes. From the table, we can see that the three involved components show inferior performance, illustrating that they play significant roles in our proposed model. noCDGCN and noCCAN obtain unsatisfactory performance, demonstrating that the categorical dynamic graph convolutional network and category-aware contextual attention network can not only adaptively capture the semantic correlations between categories but also further sufficiently aggregate multi-modal features from different aspects. noMMAN achieves better performance compared

Table 2.5 Performance comparison of the components involved in our proposed method in terms of AUC on FashionVC and ExpFashion

Method	FashionVC	ExpFashion
noCDGCN	0.7254	0.8847
noCCAN	0.7645	0.8964
noMMAN	0.7841	0.9022
FCM-CMAN	**0.8380**	**0.9115**

Table 2.6 Performance comparison of different modality combinations in terms of AUC on FashionVC and ExpFashion

Method	FashionVC	ExpFashion
Visual	0.7036	0.7319
Visual+Category	0.7645	0.8964
Textual	0.6454	0.6841
Textual+Category	0.6876	0.7097
Visual+Textual	0.7148	0.7926
All	**0.8380**	**0.9115**

to noCDGCN and noCCAN, indicating that the visual-semantic embeddings from multiple levels enhanced by the self-attention mechanism have certain characteristics to benefit fashion compatibility modeling task.

To explore the effectiveness of different modalities in our proposed model, we compared the FCM-CMAN with multiple modality combinations. Table 2.6 shows the performance comparison of different modality combinations. From the table, the following observations can be made:

- All single modalities produce positive effects on performance to varying degrees. The visual modality achieves superior performance over the textual modality, indicating that the visual modality incorporates more valuable information for compatibility modeling.
- When two random modalities are combined with each other, the performance is correspondingly improved compared with the original single modality, indicating the complementarity between different modalities.
- "Visual+Category" outperforms "Visual+Textual", indicating that category information as high-level semantics can guide the visual modality for multimodal representation learning better than the textual modality.
- When the visual and all other modalities are integrated by the contextual attention mechanism, "All" supasses other multimodal combination schemes, which indicates that the

Table 2.7 Performance comparison with different levels of visual outputs combinations in terms of AUC on FashionVC and ExpFashion

Method	FashionVC	ExpFashion
Layer 4	0.8064	0.8397
Layer 4+3	0.8177	0.8483
Layer 4+3+2	0.8231	0.8920
Layer 4+3+2+1	**0.8380**	**0.9115**

complete category-aware multimodal fusion method can better exploit complementarity among different modalities to improve model performance.

Additionally, we investigated the influence of the visual outputs from different convolutional layers, and the performance is reported in Table 2.7. From the table, we found that the output of each layer has a positive effect on our proposed model, which validates the effectiveness of the multi-layer representation fusion strategy, and the low-level to high-level visual features can provide more detailed cues.

2.4.3   Fashion Item Retrieval

To assess the practical value of our proposed compatibility modeling approach, we conducted experiments with respect to complementary fashion item retrieval in two classical outfit matching tasks, i.e., *top-to-bottom* and *bottom-to-top*. Specifically, we first picked serveral top or bottom items that appeared in S_{test} to build a query set and randomly selected 10 corresponding fashion items as the ranking candidates, including only one positive item and 9 negative items. Then, we fed them into different trained models and obtained the corresponding ranking results for each query item based on the predicted compatibility scores. Figure 2.6 shows the retrieval ranking results of our proposed FCM-CMAN and its several ablation schemes. From the figure, we found that our proposed FCM-CMAN can well retrieve positive items that are compatible with query items for different collocation problems. In the first example, only complete FCM-CMAN model identified the correct compatible item, which further demonstrates the importance of each component involved in our proposed FCM-CMAN. As for the last example, only the noCDGCN scheme made a wrong choice, but the light pink sleeveless top appeared first in the ranking results of fashion retrieval. This may be due to the similarity in style and color between the candidate item and the query item white trousers, and also implies the effectiveness of dynamic representation learning of categories and contextual attention mechanism.

Fig. 2.6 Retrieval result illustration of our proposed FCM-CMAN and its several ablation schemes for the given testing tops and bottoms. The fashion items highlighted in the red boxes are positive ones that can be used as the ground truth

2.5   Conclusion and Future Work

In this work, we propose a category-aware multimodal attention network for fashion compatibility modeling, named as FCM-CMAN. To better address issues concerning the underutilization of multimodal information and lack of semantic guidance, we sought the multimodal representations of fashion items based on category-aware visual-semantic embedding learning instead of traditional single modalities. Extensive experiments over FashionVC and ExpFashion datasets verified the superiority of our proposed FCM-CMAN, indicating the effectiveness of the categorical dynamic graph convolutional network and the category-aware contextual attention network for aggregating multimodal information. In addition, the multi-layered multi-head attention network is helpful to further capture significant elements related to fashion compatibility. One limitation of our work is that we currently ignore user's style preferences in fashion compatibility modeling. In the future, we will focus on exploring personalized and style preference-based fashion recommendation.

References

1. Baltrušaitis, T., Ahuja, C., Morency, L.P.: Multimodal machine learning: A survey and taxonomy. IEEE Transactions on Pattern Analysis and Machine Intelligence **41**(2), 423–443 (2018)
2. Chen, H., Ding, G., Liu, X., Lin, Z., Liu, J., Han, J.: Imram: Iterative matching with recurrent attention memory for cross-modal image-text retrieval. In: Proceedings of the IEEE/CVF Conference on Computer Vision and Pattern Recognition, pp. 12655–12663 (2020)

3. Chen, L., He, Y.: Dress fashionably: Learn fashion collocation with deep mixed-category metric learning. In: Proceedings of the AAAI Conference on Artificial Intelligence, pp. 2103–2110 (2018)
4. Chen, W., Huang, P., Xu, J., Guo, X., Guo, C., Sun, F., Li, C., Pfadler, A., Zhao, H., Zhao, B.: Pog: personalized outfit generation for fashion recommendation at alibaba ifashion. In: Proceedings of the International ACM SIGKDD Conference on Knowledge Discovery and Data Mining, pp. 2662–2670 (2019)
5. Chen, X., Chen, H., Xu, H., Zhang, Y., Cao, Y., Qin, Z., Zha, H.: Personalized fashion recommendation with visual explanations based on multimodal attention network: Towards visually explainable recommendation. In: Proceedings of the International ACM SIGIR Conference on Research and Development in Information Retrieval, pp. 765–774 (2019)
6. Cucurull, G., Taslakian, P., Vazquez, D.: Context-aware visual compatibility prediction. In: Proceedings of the IEEE/CVF Conference on Computer Vision and Pattern Recognition, pp. 12617–12626 (2019)
7. Cui, Z., Li, Z., Wu, S., Zhang, X.Y., Wang, L.: Dressing as a whole: Outfit compatibility learning based on node-wise graph neural networks. In: Proceedings of the ACM International Conference on World Wide Web Conference, pp. 307–317 (2019)
8. Dong, X., Song, X., Zheng, N., Wu, J., Dai, H., Nie, L.: Tryoncm2: Try-on-enhanced fashion compatibility modeling framework. IEEE Transactions on Neural Networks and Learning Systems (2022)
9. Gu, X., Wong, Y., Shou, L., Peng, P., Chen, G., Kankanhalli, M.S.: Multi-modal and multi-domain embedding learning for fashion retrieval and analysis. IEEE Transactions on Multimedia **21**(6), 1524–1537 (2018)
10. Guan, W., Wen, H., Song, X., Yeh, C.H., Chang, X., Nie, L.: Multimodal compatibility modeling via exploring the consistent and complementary correlations. In: Proceedings of the ACM International Conference on Multimedia, pp. 2299–2307 (2021)
11. Guo, B., Zhang, C., Liu, J., Ma, X.: Improving text classification with weighted word embeddings via a multi-channel textcnn model. Neurocomputing **363**, 366–374 (2019)
12. Han, X., Song, X., Yao, Y., Xu, X.S., Nie, L.: Neural compatibility modeling with probabilistic knowledge distillation. IEEE Transactions on Image Processing **29**, 871–882 (2019)
13. Han, X., Song, X., Yin, J., Wang, Y., Nie, L.: Prototype-guided attribute-wise interpretable scheme for clothing matching. In: Proceedings of the International ACM SIGIR Conference on Research and Development in Information Retrieval, pp. 785–794 (2019)
14. Han, X., Wu, Z., Jiang, Y.G., Davis, L.S.: Learning fashion compatibility with bidirectional LSTMs. In: Proceedings of the ACM International Conference on Multimedia, pp. 1078–1086 (2017)
15. He, K., Zhang, X., Ren, S., Sun, J.: Deep residual learning for image recognition. In: Proceedings of the IEEE/CVF Conference on Computer Vision and Pattern Recognition, pp. 770–778 (2016)
16. Huang, F., Zhang, X., Xu, J., Zhao, Z., Li, Z.: Multimodal learning of social image representation by exploiting social relations. IEEE Transactions on Cybernetics **51**(3), 1506–1518 (2019)
17. Jing, P., Ye, S., Nie, L., Liu, J., Su, Y.: Low-rank regularized multi-representation learning for fashion compatibility prediction. IEEE Transactions on Multimedia **22**(6), 1555–1566 (2019)
18. Jing, P., Zhang, J., Nie, L., Ye, S., Liu, J., Su, Y.: Tripartite graph regularized latent low-rank representation for fashion compatibility prediction. IEEE Transactions on Multimedia **24**, 1277–1287 (2021)
19. Kaicheng, P., Xingxing, Z., Wong, W.K.: Modeling fashion compatibility with explanation by using bidirectional lstm. In: Proceedings of the IEEE/CVF Conference on Computer Vision and Pattern Recognition, pp. 3894–3898 (2021)
20. Kingma, D.P., Ba, J.: Adam: A method for stochastic optimization. In: Proceedings of the International Conference on Learning Representations, pp. 1–15 (2014)

21. Kuang, Z., Gao, Y., Li, G., Luo, P., Chen, Y., Lin, L., Zhang, W.: Fashion retrieval via graph reasoning networks on a similarity pyramid. In: Proceedings of the IEEE/CVF International Conference on Computer Vision, pp. 3066–3075 (2019)

22. Li, X., Wang, X., He, X., Chen, L., Xiao, J., Chua, T.S.: Hierarchical fashion graph network for personalized outfit recommendation. In: Proceedings of the International ACM SIGIR Conference on Research and Development in Information Retrieval, pp. 159–168 (2020)

23. Li, Y., Cao, L., Zhu, J., Luo, J.: Mining fashion outfit composition using an end-to-end deep learning approach on set data. IEEE Transactions on Multimedia **19**(8), 1946–1955 (2017)

24. Lin, M., Chen, Q., Yan, S.: Network in network. arXiv:1312.4400 (2013)

25. Lin, Y., Ren, P., Chen, Z., Ren, Z., Ma, J., De Rijke, M.: Explainable outfit recommendation with joint outfit matching and comment generation. IEEE Transactions on Knowledge and Data Engineering **32**(8), 1502–1516 (2019)

26. Lin, Y.L., Tran, S., Davis, L.S.: Fashion outfit complementary item retrieval. In: Proceedings of the IEEE/CVF Conference on Computer Vision and Pattern Recognition, pp. 3311–3319 (2020)

27. Liu, X., Sun, Y., Liu, Z., Lin, D.: Learning diverse fashion collocation by neural graph filtering. IEEE Transactions on Multimedia **23**, 2894–2901 (2020)

28. Lu, Z., Hu, Y., Chen, Y., Zeng, B.: Personalized outfit recommendation with learnable anchors. In: Proceedings of the IEEE/CVF Conference on Computer Vision and Pattern Recognition, pp. 12722–12731 (2021)

29. McAuley, J., Targett, C., Shi, Q., Van Den Hengel, A.: Image-based recommendations on styles and substitutes. In: Proceedings of the International ACM SIGIR Conference on Research and Development in Information Retrieval, pp. 43–52 (2015)

30. Mikolov, T., Sutskever, I., Chen, K., Corrado, G.S., Dean, J.: Distributed representations of words and phrases and their compositionality. In: Proceedings of the Advances in Neural Information Processing Systems, pp. 3111–3119 (2013)

31. Peng, Y., Qi, J., Yuan, Y.: Modality-specific cross-modal similarity measurement with recurrent attention network. IEEE Transactions on Image Processing **27**(11), 5585–5599 (2018)

32. Pennington, J., Socher, R., Manning, C.D.: Glove: Global vectors for word representation. In: Proceedings of the Conference on Empirical Methods in Natural Language Processing, pp. 1532–1543 (2014)

33. Ramachandram, D., Taylor, G.W.: Deep multimodal learning: A survey on recent advances and trends. IEEE Signal Processing Magazine **34**(6), 96–108 (2017)

34. Rendle, S., Freudenthaler, C., Gantner, Z., Schmidt-Thieme, L.: BPR: bayesian personalized ranking from implicit feedback. In: Proceedings of the Conference on Uncertainty in Artificial Intelligence, pp. 452–461. AUAI Press (2009)

35. Severyn, A., Moschitti, A.: Twitter sentiment analysis with deep convolutional neural networks. In: Proceedings of the International ACM SIGIR Conference on Research and Development in Information Retrieval, pp. 959–962 (2015)

36. Shi, J., Zhang, H., Zhou, D., Zhang, Z.: Toward intelligent interactive design: A generation framework based on cross-domain fashion elements. In: Proceedings of the ACM International Conference on Multimedia (2023)

37. Song, X., Fang, S.T., Chen, X., Wei, Y., Zhao, Z., Nie, L.: Modality-oriented graph learning toward outfit compatibility modeling. IEEE Transactions on Multimedia (2021)

38. Song, X., Feng, F., Han, X., Yang, X., Liu, W., Nie, L.: Neural compatibility modeling with attentive knowledge distillation. In: Proceedings of the International ACM SIGIR Conference on Research and Development in Information Retrieval, pp. 5–14 (2018)

39. Song, X., Feng, F., Liu, J., Li, Z., Nie, L., Ma, J.: Neurostylist: Neural compatibility modeling for clothing matching. In: Proceedings of the ACM International Conference on Multimedia, pp. 753–761 (2017)

40. Song, X., Han, X., Li, Y., Chen, J., Xu, X.S., Nie, L.: Gp-bpr: Personalized compatibility modeling for clothing matching. In: Proceedings of the ACM International Conference on Multimedia, pp. 320–328 (2019)

41. Tan, R., Vasileva, M.I., Saenko, K., Plummer, B.A.: Learning similarity conditions without explicit supervision. In: Proceedings of the IEEE/CVF International Conference on Computer Vision, pp. 10373–10382 (2019)

42. Vasileva, M.I., Plummer, B.A., Dusad, K., Rajpal, S., Kumar, R., Forsyth, D.: Learning type-aware embeddings for fashion compatibility. In: Proceedings of the European Conference on Computer Vision, pp. 390–405 (2018)

43. Vaswani, A., Shazeer, N., Parmar, N., Uszkoreit, J., Jones, L., Gomez, A.N., Kaiser, Ł., Polo-sukhin, I.: Attention is all you need. In: Proceedings of the Advances in Neural Information Processing Systems, pp. 6000–6010 (2017)

44. Wang, D., Cui, P., Ou, M., Zhu, W.: Learning compact hash codes for multimodal representations using orthogonal deep structure. IEEE Transactions on Multimedia 17(9), 1404–1416 (2015)

45. Wang, X., Wu, B., Zhong, Y.: Outfit compatibility prediction and diagnosis with multi-layered comparison network. In: Proceedings of the ACM International Conference on Multimedia, pp. 329–337 (2019)

46. Wu, Z., Zheng, C., Cai, Y., Chen, J., Leung, H.f., Li, Q.: Multimodal representation with embed-ded visual guiding objects for named entity recognition in social media posts. In: Proceedings of the ACM International Conference on Multimedia, pp. 1038–1046 (2020)

47. Yang, X., Ma, Y., Liao, L., Wang, M., Chua, T.S.: Transnfcm: Translation-based neural fashion compatibility modeling. In: Proceedings of the AAAI Conference on Artificial Intelligence, pp. 403–410 (2019)

48. Yang, X., Ramesh, P., Chitta, R., Madhvanath, S., Bernal, E.A., Luo, J.: Deep multimodal rep-resentation learning from temporal data. In: Proceedings of the IEEE/CVF Conference on Com-puter Vision and Pattern Recognition, pp. 5447–5455 (2017)

49. Yang, X., Xie, D., Wang, X., Yuan, J., Ding, W., Yan, P.: Learning tuple compatibility for conditional outfit recommendation. In: Proceedings of the ACM International Conference on Multimedia, pp. 2636–2644 (2020)

50. Yu, Z., Yu, J., Fan, J., Tao, D.: Multi-modal factorized bilinear pooling with co-attention learning for visual question answering. In: Proceedings of the IEEE/CVF International Conference on Computer Vision, pp. 1821–1830 (2017)

51. Yue, L., Zhou, D., Xie, L., Zhang, F., Yan, Y., Yin, E.: Safe-vln: Collision avoidance for vision-and-language navigation of autonomous robots operating in continuous environments. IEEE Robotics and Automation Letters (2024)

52. Zhan, H., Lin, J., Ak, K.E., Shi, B., Duan, L.Y., Kot, A.C.: A3-fkg: Attentive attribute-aware fashion knowledge graph for outfit preference prediction. IEEE Transactions on Multimedia 24, 819–831 (2021)

53. Zhang, H., Zha, Z.J., Yang, Y., Yan, S., Gao, Y., Chua, T.S.: Attribute-augmented semantic hierarchy: towards bridging semantic gap and intention gap in image retrieval. In: Proceedings of the ACM International Conference on Multimedia, pp. 33–42 (2013)

Try-On-Enhanced Fashion Compatibility Modeling

3

3.1 Introduction

In this chapter, we delve into the automatic modeling of fashion compatibility using a coarse-grained Tryon-enhanced scheme. The recent expansion of the e-commerce fashion industry has exponentially increased the availability of various fashion items.[1] It not only brings consumers more choices, but also makes it more difficult for them to compose compatible items (e.g., a T-shirt and a jeans) and make outfits from the tremendous fashion item market. To address this practical issue, increasing research efforts [7, 14, 16, 34] have been paid to the fashion compatibility modeling, which aims to automatically evaluate the matching degree of an outfit (i.e., a set of fashion items). Existing approaches mainly analyze the fashion compatibility based on the *discrete item interaction*, where an outfit is treated as a composition of discrete fashion items and its compatibility is measured by certain metrics over the features of these fashion items [6, 14, 34, 43]. Although these efforts have obtained promising achievements, they overlook a practical factor that affects the fashion compatibility evaluation: the *try-on appearance* of the outfit, where items would be spatially distributed according to their functions/categories and partially covered by each other. As illustrated in Fig. 3.1a, in the try-on image (i.e., the left image), the striped T-shirt is partially covered by the blazer on the left and right sides and the black leggings below. Apparently, these spatial arrangement and occlusion can be hardly reflected in their discrete item images (i.e., the images on the right side). Therefore, we argue the importance of analyzing the try-on appearance in the fashion compatibility modeling.

Towards this end, in this work, we propose to enhance the traditional discrete item interaction modeling with the combined try-on appearance modeling, which analyzes the fashion compatibility based on the outfit's try-on appearance. However, this is non-trivial due to the

---

[1] https://fashionunited.com/global-fashion-industry-statistics/.

W. Guan et al., *Advanced Multimodal Compatibility Modeling and Recommendation*,
Synthesis Lectures on Information Concepts, Retrieval, and Services,
https://doi.org/10.1007/978-3-031-81048-0_3

Fig. 3.1 Three examples of well-matched outfits, consisting of the outfit's try-on appearance image (the left image of each composition) and discrete images of its composing fashion items (the right images of each composition)

following challenges: (i) as the try-on appearance image of an outfit is not always available for testing, how to automatically derive the try-on appearance image of the outfit based on its item images is the first challenge; (ii) how to analyze the fashion compatibility from the try-on appearance image and hence promote the fashion compatibility modeling performance constitutes another tough challenge; and (iii) as only positive outfits can be obtained from the fashion-oriented websites, existing studies mainly use the random sampling strategy to compose negative outfits to train their models [7, 34, 39]. However, the "negative" outfits derived in this manner maybe not always reliable, which can be potentially positive due to the fact that some items (e.g., jeans) can go well with various kinds of items. Therefore, how to formulate the problem concentrating on the realiable positive outfits is another key challenge.

To address the challenges, based on the previous work [6], we propose a Try-on-enhanced Fashion Compatibility Modeling framework TryonCM2, which jointly models the fashion compatibility from perspectives of the discrete item interaction and combined try-on appearance. In our context, we term the fashion compatibility from the two perspectives as the discrete and combined compatibilities, respectively. As illustrated in Fig. 3.2, as for a well-matched outfit, we extract the item features based on the items' images through a CNN-based feature extractor. In the discrete item interaction modeling, we regard each outfit as an item sequence, and employ a bidirectional LSTM network to uncover the latent interaction among item by keeping predicting the next compatible item for the previous given items. Regarding the combined try-on appearance modeling, we first generate a try-on template by a template decoder based on item features to depict the outfit's try-on appearance. Then, we split the try-on template into image stripes in both horizontal and vertical directions, and similarly employ the bidirectional LSTM network to learn the contextual structure of the try-on template by keeping predicting the next stripe according to the previous ones. Ultimately, we assume that the TryonCM2 trained with only well-matched outfits can grasp the

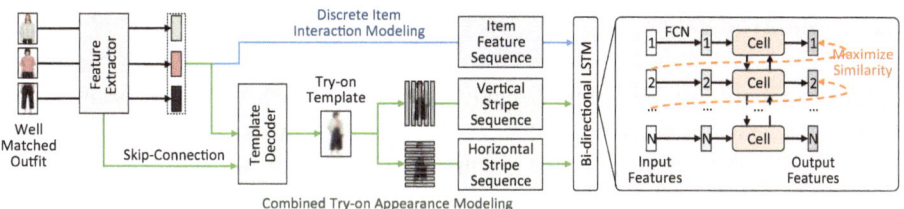

Fig. 3.2 Illustration of the proposed TryonCM2 framework, which consists of the discrete item interaction modeling (blue flow) and combined try-on appearance modeling (green flow) that model the fashion compatibility from the discrete and combined perspectives, respectively

fashion collocation patterns and can work as a "fashion prototype" to provide compatible reference outfits from either the discrete item interaction or combined try-on appearance perspectives. Therefore, we measure the fashion compatibility of a new outfit based on the similarity between this outfit and the compatible outfits produced by TryonCM2.

It is worth mentioning that, as compared with our previous work TryOn-CM [7], this latest approach has the following improvements: (i) the previous TryOn-CM encodes the outfit's combined compatibility by a single latent vector, which may miss the important clues of the try-on appearance. Differently, TryonCM2 models the combined compatibility by directly analyzing the contextual structure of the generated try-on template; (ii) the previous TryOn-CM casts the task of fashion compatibility modeling as a classification problem, where incompatible (negative) outfits need to be randomly sampled. Differently, TryonCM2 only relies on the well-matched outfits, which omits the unreliable negative sampling process; and (iii) to justify TryonCM2, we conduct more experiments including both the objective and subjective evaluations.

The main contribution can be summarized in threefold:

1. We introduce a new try-on-enhanced fashion compatibility modeling framework (TryonCM2), which jointly integrates the discrete item interaction modeling and the outfit's try-on appearance modeling. To the best of our knowledge, we are the first to explore the contextual structure of the try-on appearance to enhance the fashion compatibility modeling.
2. We propose a novel training scheme for TryonCM2, which only relies on the well-matched outfits, omitting the unreliable negative sampling process. Meanwhile, we introduce a new compatibility evaluation method, in which the compatible outfits produced by well-trained TryonCM2 can be treated as the reference.
3. Extensive objective and subjective experiments conducted on the FOTOS dataset demonstrate the portability and effectiveness of the proposed framework. We will release the codes to benefit other researchers.

3.2   Related Work

Fashion Compatibility Modeling. Recently, increasing research efforts have been dedicated to coping with the fashion compatibility modeling problem, due to its huge practical value. Existing work can be roughly divided into three groups: the pair-wise [16, 21, 34], list-wise [2, 14] and graph-based methods [4, 41]. The pair-wise method treats an outfit as multiple item pairs. For example, Song et al. [34] collected the outfit dataset from Polyvore and introduced a content-based neural framework for the compatibility modeling between the top and bottom items. Later, several studies improve the performance by involving the extra information such as the human knowledge [11, 33], and item categories [21, 24, 35]. Despite their promising results, pair-wise methods are not efficient to cope with the outfit that consists of multiple items, as it needs to compute the compatibility of each pair of items. Differently, the list-wise and graph-based methods can directly calculate the compatibility of multiple items. In particular, the list-wise method regards the outfit as an ordered sequence of fashion items and utilizes the sequential method, such as the LSTM [14, 31] and GRU [2], to uncover the sequential relationships among fashion items, while the graph-based method [4, 19, 40] focuses on constructing a graph of the fashion items and employs graph neural networks to explore the outfit compatibility. Moreover, to promote the practical value, several efforts have been made to make the compatibility more interpretable with the attention mechanism [22, 37] or interpretable feature learning [9, 12, 42].

Despite their effectiveness, these methods mostly learn the outfit compatibility based on the discrete item interaction, which ignore the combined compatibility in the outfit's try-on appearance. Therefore, in this work, we proposed to analyze the contextual structure of the try-on appearance image and incorporate it with the discrete item interaction for a better modeling of the fashion compatibility.

Deep Image Synthesis. At the initial stage, the Auto-Encoder (AE) [8] and its variants, such as the denoising AE [36] and variational AE [26] were widely used in the image synthesis task. However, most of AE-based methods utilize the mean-square loss as the loss function, resulting in the synthesized images more fuzzier. Later, Goodfellow et al. [10] proposed the Generative Adversarial Nets (GANs) and introduced an adversarial loss, which makes remarkable progress in the image synthesis task [27, 45]. After that, to incorporate the desired properties in the generated samples, researchers explored different signals, such as the text [28] and attributes [30], as priors to condition the image synthesis process. Moreover, a few studies investigated the problem of image-to-image translation via conditional GANs [17], which transforms the given input image to another one with a different representation. However, most of GANs are very deep neural networks and suffer from the information loss and degradation problem. Therefore, Ronneberger et al. [29] proposed a skip connection structure, where the encoded feature maps is delivered to its corresponding block of the decoder to merge more features during the image synthesis.

Recently, many researchers have noticed the exciting prospect of the image synthesis technology in the fashion domain, such as the fashion compatibility modeling [20], item attribute manipulation [1, 44] and virtual try-on [15, 46]. Although existing efforts have been dedicated to exploring the generative methods to enhance the compatibility modeling, they focused on generating a compatible item template for the existing ones by analyzing the discrete item interactions [23]. On the contrary, in this chapter, we planned to generate a try-on template to depict the try-on appearance of an outfit, and then analyzed the combined compatibility based on it.

3.3   Methodology

In this section, we first briefly give the definition of our fashion compatibility modeling problem. Thereafter, we present the feature encoder, and discrete item interaction modeling and combined try-on appearance modeling. Finally, we introduce how we evaluate the fashion compatibility.

3.3.1   Problem Formulation

Suppose that we have a training set $\Omega = \{(\mathbf{X}^j, \mathbf{P}^j) | j = 1, \ldots, M\}$, where $\mathbf{X}^j = [\mathbf{x}_1^j, \mathbf{x}_2^j, \ldots, \mathbf{x}_{N_j}^j]$ is the j-th outfit composed by N_j items, and $\mathbf{P}^j$ is the real try-on appearance image of the outfit $\mathbf{X}^j$. It is worth mentioning that the try-on appearance image of each outfit is only available during the training phase, used for training the try-on template synthesis. In other words, we aim to devise a comprehensive fashion compatibility modeling framework $\mathcal{F}$ simply based on the images of its composing fashion items, which is able to infer the compatibility of an outfit from perspectives of both discrete item interaction and combined try-on appearance, as follows,

$$y^j = \mathcal{F}(X^j | \Theta), \tag{3.1}$$

where Θ is a set of to-be-learned parameters. y^j denotes the compatibility score of the given outfit X^j. For simplicity, we omit the subscript j of X^j in the following illustration.

3.3.2   Feature Extractor

Due to the excellent performance of the convolutional neural network (CNN) [18] in many image understanding tasks [5, 14, 24, 47, 48], we adopt it as our feature extractor $\mathcal{E}$. Specifically, following [48], our feature extractor is devised with K convolutional layers, each of which is followed by a ReLU activation function and a batch-normalization layer. Formally, the feature extractor $\mathcal{E}$ takes the image of the item $\mathbf{x}_i$ as the input and outputs its feature vector $\mathbf{f}_i$ as follows,

$$f_i = \mathcal{E}(x_i | \Theta_{\mathcal{E}}):$$
$$\begin{cases} f_i^1 = \text{ReLU}(\text{conv}(x_i)), \\ f_i^k = \text{ReLU}(\text{bn}(\text{conv}(f_i^{k-1})))|_{k=2}^{K-1}, \\ f_i = f_i^K = \text{conv}(f_i^{K-1}), \end{cases} \tag{3.2}$$

where conv(), bn() and ReLU() refer to the convolutional layer, batch-normalization layer and ReLU activation function, respectively. $\Theta_{\mathcal{E}}$ is the set of parameters of the feature extractor. f_i^k is the feature map after the k-th convolutional layer for the input item image x_i.

3.3.3 Discrete Item Interaction Modeling

In this part, we aim to uncover the compatibility relationships in the fashion items. As same with [7, 14], we treat each outfit as an ordered sequence of fashion items, and each item in the outfit refers to a time step. Notably, these items are ordered according to their categories. Without losing generality, we pre-define the order as from the outer to top and then bottom. We then utilize a one-layer bidirectional LSTM network to comprehensively capture the underlying compatibility relationship among the image features. Here, we take the forward LSTM network as an example, while the backward LSTM can be defined similarly. The forward LSTM network recurrently takes the feature vector $\{f_i\}_{i=1}^{N}$ of each item $\{x_i\}_{i=1}^{N}$ in an outfit as the input and outputs a list of hidden vectors $\overrightarrow{h_d^i}$ as follows,

$$\overrightarrow{h_d^i} = \overrightarrow{\text{LSTM}}_d(g_d(f_i) | \Theta_d), i = 1, 2, \ldots, N, \tag{3.3}$$

where $g_d(\cdot)$ is a mapping function that transforms the feature vector into the input vector of the LSTM. In our context, following the study [14], we adopt one fully connected layer as the mapping function. Θ_d is the set of parameters of the discrete item interaction modeling. In a sense, the output $\overrightarrow{h_d^i}$ can be regarded as the predicted feature of the $(i + 1)$-th item based on the previous i input items.

To capture the interaction between item features, we expect the bidirectional LSTM network can accurately predict the next fashion item (i.e., the complementary compatible item) for the previous ones. Accordingly, we calculate the similarity between the predicted feature $\overrightarrow{h_d^i}$ and the true next feature f_{i+1} by the following objective function,

$$\mathcal{L}_d^f = \frac{1}{N-1} \sum_{i=1}^{N-1} \log \frac{\exp(\overrightarrow{h_d^i} \cdot f_{i+1})}{\sum_{f \in F} \exp(\overrightarrow{h_d^i} \cdot f)}, \tag{3.4}$$

where we adopt the inner product of the vectors as their similarity. F contains all image features of the current batch. Note that we can choose F to be the set of image features of all fashion items in the whole dataset. However, this is not practical due to the large number of items and high-dimensional representation of the image. Therefore, following [14], we set F to be all the possible features in a batch to speed up the training. By maximizing the objective function, the the predicted feature can be enforced to as similar as the true next feature, and we can learn the interaction between the fashion items in the forward direction.

Similarly, the backward LSTM network maps the feature vectors into the backward hidden vectors $\overleftarrow{h}_d^i$'s, where the items are fed in the reverse order. Accordingly, the objective function of the backward LSTM can be defined as $\mathcal{L}_d^b$ in the same form of $\mathcal{L}_d^f$ by replacing the corresponding inputs. Ultimately, the total objective function of the discrete item interaction modeling is as follows,

$$\mathcal{L}_d = \mathcal{L}_d^f + \mathcal{L}_d^b. \tag{3.5}$$

3.3.4 Combined Try-On Appearance Modeling

We aim to enhance the fashion compatibility modeling through analyzing the combined compatibility lying in the outfit's try-on appearance. Therefore, we propose to first synthesize a try-on template to depict the real try-on appearance of an outfit, and then analyze the combined try-on compatibility based on the synthesized try-on template image.

Try-on Template Synthesis. Similar to [7], we cast the task of the try-on template synthesis as an image-to-image translation problem, where the inputs are images of fashion items in the outfit and the output is the image of the try-on template. It is worth noting that we currently focus on the general compatibility modeling and thus only consider the objective factors during the template generation process, and leave out the subjective factors in our future work, such as the personal identity and body shape.

Due to the remarkable performance of the auto-encoder structure in the image-to-image translation [29, 48, 49], we adopt it to synthesize the try-on template image. Typically, in the auto-encoder structure, the encoder works on compressing the input image into a dense feature vector, while the decoder aims to translate this feature vector into the desired image. In our context, as the pre-defined feature extractor $\mathcal{E}$ can encode an arbitrary fashion item, we directly treat it as the encoder for the try-on template synthesis. Thereafter, we feed the concatenation of all the feature vectors within an outfit into the template decoder $\mathcal{D}$, which consists of K' deconvolutional layers. In addition, to reduce the information loss during the encoding process [17, 29], we incorporate the skip-connection structure into the try-on template decoder. One natural way of introducing the skip-connection is by transferring the encoded feature maps of all items in the outfit to the decoder. However, this would inevitably lead to the huge memory and time cost. Consequently, we introduce the max pooling operation before transferring the feature maps, which has achieved great success in the feature filtering [18]. Formally, the try-on template $\mathbf{P}'$ can be synthesized by the try-on template decoder $\mathcal{D}$ as follows,

$$\mathbf{P}' = \mathcal{D}(\boldsymbol{x}_1, \boldsymbol{x}_2, \ldots, \boldsymbol{x}_N | \boldsymbol{\Theta}_{\mathcal{D}}):$$

$$\begin{cases} \boldsymbol{f}_1, \boldsymbol{f}_2, \ldots, \boldsymbol{f}_N = \mathcal{E}(\boldsymbol{x}_1), \mathcal{E}(\boldsymbol{x}_2), \ldots, \mathcal{E}(\boldsymbol{x}_3), \\ \boldsymbol{h}_{\mathcal{D}}^1 = \text{ReLU}(\text{bn}(\text{deconv}(\boldsymbol{f}_1 || \boldsymbol{f}_2 || \cdots || \boldsymbol{f}_N))), \\ \boldsymbol{h}_{\mathcal{D}}^k = \text{ReLU}(\text{bn}(\text{deconv}(\boldsymbol{h}_{\mathcal{D}}^{k-1} || \boldsymbol{f}_{max}^{K-k+1})))|_{k=2}^{K'-1}, \\ \mathbf{P}' = \boldsymbol{h}_i^{K'} = \tanh(\text{deconv}(\boldsymbol{h}_i^{K'-1} || \boldsymbol{f}_{max}^{K-K'+1})), \end{cases} \tag{3.6}$$

where $\cdot || \cdot$ is the concatenation of feature maps in the last dimension. deconv() and tanh() are the deconvolutional layer and tanh activation function, respectively. $\boldsymbol{f}_{max}^k$ is the max pooling result of the feature maps $\{\boldsymbol{f}_i^k\}_{i=1}^N$ derived from the k-th convolutional layer defined in Eq. (3.2). $\boldsymbol{\Theta}_{\mathcal{D}}$ is the set of parameters of the template decoder.

To regularize the synthesized image to imitate the real try-on appearance, we adopt the L_1 norm at the pixel level between the ground truth try-on image $\mathbf{P}$ and the generated try-on template $\mathbf{P}'$. Accordingly, the objective function of the try-on template decoder $\mathcal{D}$ can be defined as follows,

$$\mathcal{L}_{\mathcal{D}} = ||\mathbf{P} - \mathbf{P}'||_1, \tag{3.7}$$

where $|| \cdot ||_1$ is the L_1 norm. It is worth noting that we adopt the L_1 norm rather than the L_2 norm, as the L_1 norm encourages the synthesized image to be less blurring [17].

Combined Compatibility Analysis. As the outfit's try-on template has been synthesized, we can learn the fashion compatibility from the combined try-on perspective. Inspired by that learning the contextual clues of the local contents have proven to be effective in the image understanding [13, 38], we split the try-on template of an outfit into multiple image stripes with the fixed length in both the horizontal and vertical directions, where each image stripe brings the local contents of the try-on appearance. Then, we adopt a bidirectional LSTM network on the horizontal and vertical image stripes, respectively, to learn the intrinsic fine-grained compatibility in the try-on template.

We take the horizontal forward LSTM network to capture the relationship of image stripes as an example in Fig. 3.3, while the backward LSTM network and the vertical bidirectional LSTM network can be defined similarly. We first equally split the input $W * H$ RGB try-on template $\mathbf{P}'$ in the horizontal direction with a fixed length a ($W \bmod a = 0$ and $\frac{W}{a} \geq 2$) into multiple image stripes $\{\boldsymbol{s}_h^i \in \mathbb{R}^{a \times H \times 3} | i = 1, \ldots, \frac{W}{a}\}$. Here, we take $a = 16$ as an example

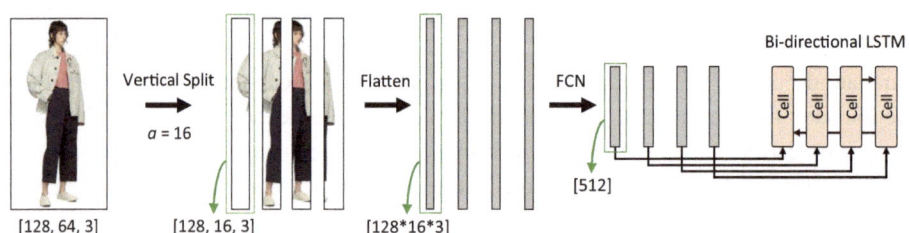

Fig. 3.3 Illustration of the horizontal forward LSTM network for the combined compatibility modeling. Here, we take the stripe length $a = 16$ as an example

in Fig. 3.3. Then, the forward horizontal LSTM takes the sequence of the image stripe from the left to the right side as the input, and outputs a sequence of hidden vectors $\overrightarrow{h}_h^i$ as follows,

$$\overrightarrow{h}_h^i = \overrightarrow{\text{LSTM}_h}(g_h(s_h^i)|\Theta_t), i = 1, 2, \ldots, \frac{W}{a}, \tag{3.8}$$

where $g_h(\cdot)$ is a fully connected layer, used for extracting the local feature of the image stripe. Θ_t is the set of parameters of the combined try-on compatibility modeling.

Algorithm 3.1 Try-on-enhanced Fashion Compatibility Modeling.

Input: The training set $\mathcal{X}$ of well-matched outfits, and total number of the propagation steps N_s.
Output: The parameter $\Theta = \{\Theta_{\mathcal{E}}, \Theta_{\mathcal{D}}, \Theta_d, \Theta_t\}$.
1: Initialize the parameters Θ.
2: **for** step $< N_s$ **do**
3:    Randomly draw batch of outfits from $\mathcal{X}$.
4:    Calculate $\mathcal{L}_d$, $\mathcal{L}_t$, and $\mathcal{L}_{\mathcal{D}}$ according to Eqs. (3.5), (3.10), and (3.7), respectively.
5:    Update the parameters:

$$\Theta_{\mathcal{E}}, \Theta_d, \Theta_{\mathcal{D}} \leftarrow \Theta_{\mathcal{E}}, \Theta_d, \Theta_{\mathcal{D}} - \eta \frac{\partial(\mathcal{L}_d + \mathcal{L}_{\mathcal{D}})}{\partial\{\Theta_{\mathcal{E}}, \Theta_d, \Theta_{\mathcal{D}}\}},$$
$$\Theta_t \leftarrow \Theta_t - \eta \frac{\partial \mathcal{L}_t}{\partial \Theta_t}.$$

6: **end for**

To capture the contextual structure of the try-on template, we expect the LSTM network can predict the next image stripe seen the previous ones. Accordingly, we maximize the similarity between the output of each time step and the corresponding ground-truth stripe feature through the following objective function,

$$\mathcal{L}_h^f = \frac{1}{N} \sum_{i=1}^{\frac{W}{a}} \log \frac{\exp(\overrightarrow{h}_h^i \cdot s_h^{i+1})}{\sum_{s_h \in S} \exp(\overrightarrow{h}_h^i \cdot s_h)}, \tag{3.9}$$

where S contains all image stripes of items in current batch.

Similarly, the backward LSTM in the horizontal direction can be defined by feeding the input in a reverse order and outputs a sequence of hidden vectors $\overleftarrow{h}_h^i$. Its loss function $\mathcal{L}_h^b$ can be also defined according to Eq.(9) by replacing the corresponding inputs. In the same way, we define the loss functions for the vertical forward and backward LSTMs as $\mathcal{L}_v^f$ and $\mathcal{L}_v^b$, respectively. Ultimately, the objective function of the combined compatibility modeling can be defined as follows,

$$\mathcal{L}_t = \mathcal{L}_h^f + \mathcal{L}_h^b + \mathcal{L}_v^f + \mathcal{L}_v^b. \tag{3.10}$$

3.3.5 Fashion Compatibility Analysis

Ultimately, the total objective function of try-on-enhanced fashion compatibility modeling can be written as follow,

$$\max_{\Theta} \mathcal{L} = \max_{\Theta}(\mathcal{L}_d + \mathcal{L}_t - \mathcal{L}_{\mathcal{D}}), \tag{3.11}$$

where $\mathcal{L}_d$, $\mathcal{L}_t$ and $\mathcal{L}_{\mathcal{D}}$ are the objective functions of the discrete item interaction modeling, combined compatibility modeling, and the try-on template synthesis defined in Eqs. (3.5), (3.10) and (3.7), respectively. $\Theta = \{\Theta_{\mathcal{E}}, \Theta_{\mathcal{D}}, \Theta_d, \Theta_t\}$ is the set of all model parameters. Algorithm 3.1 details the training process of TryonCM2. It is worth noting that parameters of the combined compatibility modeling, i.e., Θ_t, are updated separately from parameters of the template synthesizing. This is because the combined compatibility modeling is to analyze the contextual structure relationships within the try-on template and should not affect the synthesizing process.

In a sense, once our proposed framework get trained with the well-matched (positive) outfits, it can learn their fashion collocation patterns and work as a "fashion prototype" to provide compatible reference outfits from either the discrete item interaction or combined try-on appearance perspectives. Then, we regard the similarity between a new given outfit and the predicted compatible outfits as the fashion compatibility of the given outfit. Formally, we define the fashion compatibility of a given outfit $\mathbf{X}$ as follows,

$$\hat{y} = \lambda \mathcal{L}_d(\mathbf{X}) + (1 - \lambda)\mathcal{L}_t(\mathbf{X}), \tag{3.12}$$

where $\mathcal{L}_d(\mathbf{X})$ and $\mathcal{L}_t(\mathbf{X})$, defined in Eqs. (3.5) and (3.10), can be regarded as the outfit's discrete and combined compatibility, respectively. λ is a trade-off parameter to balance the two terms. For ease of illustration, we take the discrete compatibility in the forward direction as an example in Fig. 3.4. Suppose the given outfit is composed by three fashion items, i.e., $\mathbf{X} = [x_1, x_2, x_3]$. At the first time step, the forward LSTM takes the feature vector f_1 of

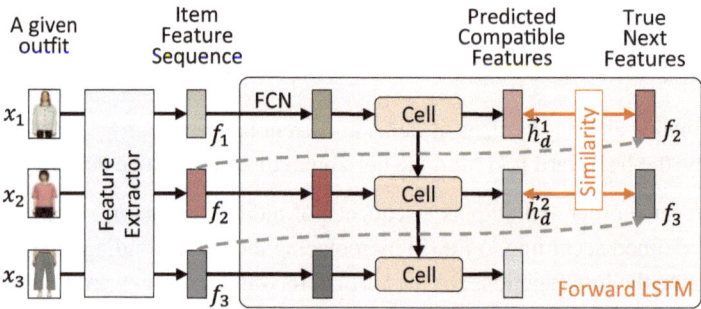

Fig. 3.4 Example of the fashion compatibility analysis in the forward LSTM of the discrete item interaction modeling

the item x_1 as the input, and predicts a compatible item feature to x_1 as $\overrightarrow{h}_d^1$. The similarity between the predicted compatible item and the true next item in the test outfit can be treated as the compatibility of x_1 and x_2. Then, at the second time step, the forward LSTM yields $\overrightarrow{h}_d^2$, indicating the feature vector of a compatible item to items x_1 and x_2. Then, the similarity of $\overrightarrow{h}_d^2$ and f_3 is regarded as the compatibility of x_3 to the previous two items. Ultimately, the discrete compatibility of the forward direction can be measured by the summation of the two similarities above, i.e., $\mathcal{L}_d^f(\mathbf{X})$.

3.4 Experiments

In this section, we gave an introduction of the dataset, detailed the network structure, and experimental settings in Sect. 3.4.1. And then, we conducted extensive experiments by answering the following research questions:

1. Does our TryonCM2 outperform the state-of-the-arts?
2. How do the hyper-parameters affect the performance?
3. How is the practical value at compatible item retrieval?
4. Are the retrieved items limited by diversity?
5. How about the synthesis quality of the try-on template?

3.4.1 Experiment Settings

Datasets. Although there are several datasets supporting the fashion compatibility modeling, like FashionVC [34] and Polyvore [14], they are not applicable for our method due to the lack of the outfit's try-on appearance image. Therefore, in this work, we chose FOTOS dataset [7], which is the only public dataset where each outfit is associated with its real try-on image. In particular, the FOTOS dataset consists of 11,000 well-matched outfits (divided into 9,900, 100, and 1,000 outfits for training, validation, and testing, respectively) composed by 20,318 fashion items. In this work, as the same with [7], we only focused on the clothing items and the outfits with no more than 4 items. And, the size of images is unified to 256×256 before being fed into the network.

 Network Structure. (i) The feature extractor consists of 8 convolutional layers, each of which involves several 5×5 spatial filters with a stride of 2. The number of filters for these layers are 64, 128, 256, 512, 512, 512, 512 and 512, respectively. (ii) The template decoder $\mathcal{D}$ composes of 7 deconvolutional layers, where 5×5 spatial filters are adopted with a stride of 1/2. The number of filters of these layers are 512, 512, 512, 256, 128, 64 and 3, respectively. Apparently, the size of the feature maps is doubled after each deconvolutional (up-sampling) layer, leading the intermediate feature vector to be decoded into the try-on template with the size of $2^7 \times 2^7 = 128 \times 128$. Due to the fact that all the outfit content of the try-on image is in the middle part with the right and left sides are blank, we cropped the template image with the size of 64×128. (iii) The bidirectional LSTM networks of the discrete item interaction modeling and combined try-on appearance modeling share the same structure, and the number of the hidden units of both networks are set to 512.

Training Setting. We used the random normal initialization for the parameters and trained them by Adam optimizer with a learning rate η as $2e^{-4}$, the dropout rate as 0.5, and batch size as 32. We utilized the grid search to tune the hyper-parameters of our method. In particular, we set the trade-off parameter λ in Eq. (3.12) as 0.7 and the length of image stripe a in Eq. (3.8) as 16. The discussion of the hyper-parameters is detailed in Sect. 3.4.3. We trained the network with 10,000 propagation steps and selected the parameters according to the performance of the validation set.

3.4.2 Comparison of Approaches

We evaluated the performance on the top-n recommendation task [3]. For each testing outfit, we randomly picked one composing fashion item as the ground truth item and randomly sampled additional 499 fashion items from the same category of the ground truth item as the candidate items. All candidate items are ranked based on their compatibility scores derived by Eq. (3.12). We adopted the Mean Reciprocal Ranking (MRR), Hit Rate (HR) and Normalized Discounted Cumulative Gain (NDCG) at 10, 50, 100 to evaluate our model in the top-n recommendation task [3, 21]. Meanwhile, we adopted the Area Under Curve (AUC) as another metric to evaluate the model in the positive/negative outfit classification.

To prove the effectiveness of the proposed TryonCM2, we chose different kinds of baselines ranging from the naive, pair-wise, list-wise to the graph-based methods.

- **RAND**. We randomly ranked the candidate items for the query fashion items and reported the results.
- **PopRank**. We ranked the candidate items directly based on its popularity, which is defined by the number of occurrences of the fashion item in the dataset.
- **BPR-DAE** [34]. This method models the coherent relation between multiple modalities, i.e., images and texts, of fashion items via an auto-encoder network.
- **Bi-LSTM-VSE** [14]. It utilizes a bidirectional LSTM to learn the discrete item interaction and design a visual-semantic space to capture the consistency of the images and texts.
- **NCR** [2]. This work makes full use of the textual information of fashion items and models the outfit compatibility from both semantic and lexical aspects.
- **PAICM** [12]. The method involves matrix factorization to learn some compatible and incompatible outfit prototypes, which can be used to guide the outfit compatibility modeling. Note here we directly utilized the pre-trained attribute predictor for our dataset following their work.

- **CSA-Net** [21]. This work presents a category-based attention selection mechanism for the complementary item retrieval problem, which utilizes an attention machine to learn the latent embeddings of items and calculates the average distance between the candidate item and each query item as the fashion compatibility.
- **NGNN**. [4] This method represents an outfit as a graph and improved a graph neural networks, where the nodes are categories and edges are the interaction between them, to analyze the relationship between fashion items.
- **HFGN** [19]. To fit the context of general fashion compatibility model in this work, following this work, we adopted a hierarchical graph upon the outfit-item interactions and aggregate the information from the item level into the outfit level to learn their representations.

Note that the above state-of-the-arts were trained based on the FOTOS dataset anew. The comparison results are listed in Table 3.1, where different kinds of methods, i.e., the naive method, pair-wise, list-wise, graph-based, and our proposed methods, are separated by lines. From the results, we have the following observations.

1. Our proposed TryonCM2 achieves the best performance with respect to all metrics, which proves the superiority of our proposed method. Besides, TryonCM2 outperforms many of other methods that utilize the multi-modal data, e.g., BPR-DAE and Bi-LSTM-VSE, although it only takes the visual information of fashion items. It may be because that TryonCM2 fully explores the visual information from both the discrete and combined perspectives.
2. On average, the list-wise methods outperform the pair-wise methods. One possible explanation is that the pair-wise methods mainly focus on the compatibility modeling between two items and fail to capture the interaction of multiple fashion items.
3. The graph-based method HFGN outperforms NGNN. This might because that HFGN enhances the item embedding with the compatibility information instead of the neighbor information, it can better model the underlying compatibilty interactions. However, HFGN performs worse than TryOn-CM that incorporates the combined try-on appearance into the list-wise item interaction modeling, which reflects the importance of the try-on appearance in the fashion compatibility modeling.

3.4.3 Hyper-Parameter Discussion

We discussed the hyper-parameters, i.e., the trade-off parameter λ in Eq. (3.12) and length of image stripe a in Eq. (3.8). **The Trade-off Parameter** λ. This hyper-parameter balances the importance of the discrete and combined compatibilities for the fashion compatibility modeling. Referring to Eq. (3.12), λ means importance of the discrete compatibility.

Table 3.1 Performance comparison among different methods, including the state-of-the-arts (the naive (i.e., RAND and PopRank), pair-wise (i.e., BPR-DAE [34], PAICM [12], and CSA-Net [21]), list-wise (i.e., NCR [2], Bi-LSTM-VSE [14], and TryOn-CM [7]), graph-based (i.e., NGNN [4] and HFGN [19]) methods) and the derivation of our method. The input of each method is illustrated correspondingly

Method	Input	AUC	MRR	HR@k			NDCG@k		
				$k = 10$	$k = 50$	$k = 100$	$k = 10$	$k = 50$	$k = 100$
RAND	–	0.502	0.014	0.020	0.100	0.204	0.008	0.025	0.041
PopRank	–	0.496	0.016	0.024	0.112	0.221	0.009	0.026	0.041
BPR-DAE [34]	Image+Text	0.742	0.087	0.165	0.394	0.552	0.087	0.137	0.162
PAICM [12]	Image	0.692	0.057	0.110	0.299	0.468	0.052	0.094	0.119
CSA-Net [21]	Image+Category	0.701	0.061	0.107	0.322	0.486	0.064	0.106	0.137
NCR [2]	Text	0.646	0.034	0.064	0.222	0.376	0.029	0.062	0.087
Bi-LSTM-VSE [14]	Image+Text	0.794	0.118	0.226	0.484	0.642	0.120	0.177	0.203
TryOn-CM [7]	Image+Text	0.832	0.134	0.290	0.582	0.721	0.154	0.220	0.242
NGNN [4]	Image+Category	0.698	0.055	0.102	0.320	0.478	0.054	0.101	0.126
HFGN [19]	Image	0.820	0.160	0.304	0.552	0.708	0.158	0.217	0.238
TryonCM2	Image	**0.873**	**0.303**	**0.464**	**0.680**	**0.778**	**0.331**	**0.379**	**0.395**

To examine the effect of the trade-off parameter λ, we reported the AUC performance of our TryonCM2 with respect to λ from 0 to 1 with a stride of 0.1 in Fig. 3.5a.

As we can see, along with the importance parameter of the discrete compatibility goes from 0 to 1, the model performance first keeps increasing until achieves the best performance when $\lambda = 0.7$, and then decreases with λ continuously increases to 1. On one hand, this suggests the necessity of jointly taking both discrete and combined compatibilities into account in the fashion compatibility modeling task. On the other hand, this reflects that the discrete compatibility modeling is more powerful than the combined compatibility modeling, and dominants the final evaluation. One possible explanation is that the combined compatibility is as an auxiliary compatibility to the discrete compatibility for the purpose of capturing the fashion compatibility from the try-on appearance.

The Length of Image Stripe a. This hyper-parameter controls the size of each image stripe when splitting the try-on template. A larger a indicates that the stripe covers more pixels of the image; more information of the local contents, and shorter length of image stripe sequence. To examined the effect of the length of image stripes a on the model performance, we set the length a of each image stripe to 2, 4, 8, 16 and 32 to split the 64×128 try-on template image, which leads the length the image stripe sequence to 32, 16, 8, 4 and 2 for the horizontal LSTM network and 64, 32, 16, 8 and 4 for the vertical LSTM network,

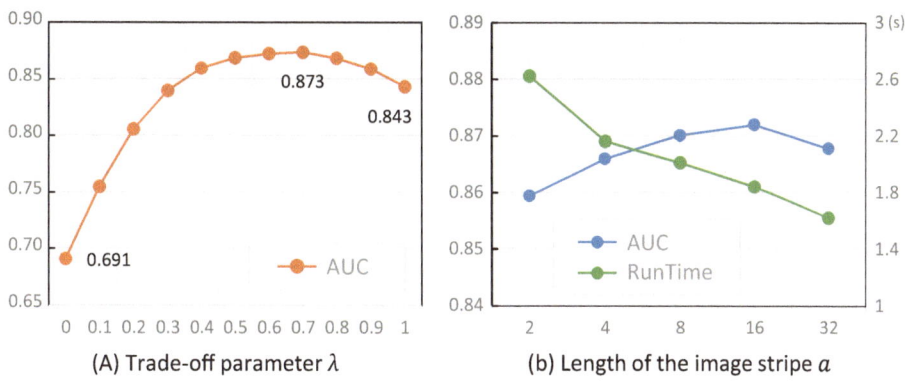

Fig. 3.5 Performance of TryonCM2 with respect to the different trade-off parameter λ in (**a**) and the number of the length a in (**b**)

respectively. In Fig. 3.5b, we reported the AUC results to examine the model accuracy, and the running time to show the model efficiency.

As we can seen, the AUC first increases to the best performance at $a = 16$, and then decreases along with the increasing of the length a. The possible reasons are as follows. When a is too small, each image stripe cannot carry enough local contents so that the LSTM networks fail to capture the contextual relationship of the try-on template image. Therefore, along with a becoming large, each stripe carries more contents and the performance increases. But when a is becoming too large, each image stripe is "pump" that carries too much local contents, so that LSTM can not capture the contextual relationship among local contents. RunTime in Fig. 3.5b refers to the time cost for testing 100 outfits. It decreases along with the sequence length of image stripe increasing. This may be because that along with a increasing, the sequence length of image stripe decreases and the time steps of the LSTM network decrease, whereby the time cost of the network saves.

3.4.4 Compatible Item Retrieval

To assess the practical value of the proposed TryonCM2 method, we conducted experiments on the visualization and subjective psycho-visual test of the complementary fashion item retrieval.

Compatible Item Retrieval Results. Without losing generality, we compared the retrieval results of our proposed TryonCM2 with strongest baseline TryOn-CM [7] in Fig. 3.6. The "Query" column shows the query item(s), and the ground-truth target item in the black circle. The right column lists the corresponding top 10 retrieval results for each query. The green check-mark indicates that the ground-truth item is successfully retrieved by the method, where the ground-truth item is surrounded by the green box, while the

Fig. 3.6 Example retrieval results of TryonCM2 and TryOn-CM. The "Query" column shows the retrieval query item(s) and the ground-truth item is surrounded by green box. The rest items in each row are the retrieved top-10 compatible items from 500 candidate items. The green check-mark and red cross-mark indicate the success and failure of the example, respectively

red cross-mark indicates the ground-truth item is unsuccessfully retrieved. Regarding the retrieval results, we have the following observations.

1. Overall, TryonCM2 is able to retrieve the ground-truth item and rank it at the front place compared with TryOn-CM. This may be because TryonCM2 learns the contextual structure in the try-on image, whereby has a better understanding of the outfit's combined compatibility. In addition, most of the rest retrieved items are also compatible to the query item(s), such as the jeans (the third, sixth, and seventh items in the first row of Fig. 3.6) can pair up with the query denim jacket and T-shirt as a casual look. And the black pants (the fifth and eighth items) pair up with the query items as a boyish style.
2. TryonCM2 sometimes retrieves the items that are visually compatible but practically incompatible to the query item(s). For example, as for the fourth example in Fig. 3.6, the shorts (the first item) and skirts (the sixth item) retrieved by TryonCM2 look well paired up with the query fur and sweatshirt, while they are incompatible with given items regarding the applicable season (i.e., the summer clothes can hardly go well with the winter ones). This may be due to that TryonCM2 is totally visual-oriented. Contrarily, TryOn-CM that involves the item's textual descriptions during fashion compatibility modeling performs better in this case.
3. In certain cases, although TryonCM2 fails to retrieve the ground-truth items, it can retrieve other items that are in the similar colors/styles with the ground-truth items. Taking the fifth query in Fig. 3.6 as an example, the top-three retrieved items by TryonCM2 are in the same color with the ground-truth white vest, which is not ranked at the top 10 places. This may be because that the query item(s) is(are) easy-matched, so that many other fashion items can pair up with it as a fashion looking. As for such cases, TryOn-CM also fails to retrieval the ground-truth compatible item.

Subjective Psycho-visual Test. To gain a more comprehensive insight of the compatible item retrieval, we randomly selected 15 queries and the corresponding most compatible item retrieved be TryOn-CM and TryonCM2. Accordingly, we totally have 15 pairs of outfits, each of which consists of an outfit composed by TryOn-CM and TryonCM2. We invited volunteers to judge which outfit in each outfit pair looks better. To facilitate the evaluation, the volunteer can choose a judgment from "The first outfit", "The second outfit", "Both", and "None". Notably, volunteers did not know which outfit is composed by TryOn-CM and TryonCM2. To ensure the collected judgments are general and meaningful for the public, we invited ordinary people as volunteers instead of fashion experts. Ultimately, we totally collected 84 results of the judgments. The information statistics of volunteers and survey results are shown in Fig. 3.7a and b, respectively. From the information statistics of volunteers, we can see that males and females of volunteers are $0.37 : 0.63$ and they are aged from 18 to 47, which ensure that the results are subjective and representative. From the judgment results we can see that 43.28% volunteers thought the outfit composed by TryonCM2 are more compatible, which has an improvement of 13.09% compared with that by TryOn-CM. This further proves that superiority of the new proposed TryonCM2.

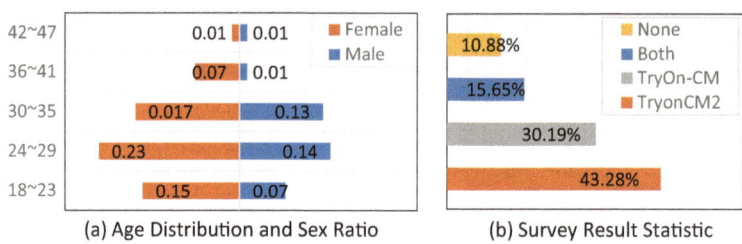

(a) Age Distribution and Sex Ratio   (b) Survey Result Statistic

Fig. 3.7 The information statistics of volunteers (**a**) and judgment results (**b**)

3.4.5 Diversity of the Retrieved Items

To examine whether the retrieved items by TryonCM2 are limited by diversity, we first quantitatively compared the retrieval diversity of TryonCM2 with the strongest baseline TryOn-CM, and then, visualized the retrieval diversity.

   Retrieval Diversity Comparison. We adopted two metrics to evaluate the average diversity of the retrieval results following the approach [25]: the intra-list distance (ILD) and F-measure (F_m), as follows,

$$ILD(k) = \frac{1}{k(k-1)} \sum_{i \neq j \in L_k} (1 - v_i \cdot v_j),$$

$$F_m(k) = \frac{2 \times NDCG(k) \times ILD(k)}{NDCG(k) + ILD(k)},$$

where L_k is recommended item list at trial k. $v_i \cdot v_j$ denotes the similarity between the items i and j in L_k. Here v_i are the VGG features [32] of the item i. $F_m(k)$ is to evaluate the trading-off between the NDCG(k) and ILD(k). A higher value of ILD (F-measure) indicates the better performance. Without losing generality, we reported ILD and F-measure at $k = 10, 50, 100$ of our method and the strongest baseline TryOn-CM [7].

   As we can see, the TryonCM2 achieves the similar performance as TryOn-CM regarding the ILD metric, while outperforms TryOn-CM regarding F-measure metric. This illustrates that TryonCM2 has improved the retrieval accuracy, while maintaining the diversity. The possible reasons are twofold. On the one hand, TryonCM2 leverages the contextual structure of the try-on template to better analyze the outfit's combined compatibility, whereby achieves the higher accuracy. On the other hand, TryOn-CM is trained with both the positive and randomly selected negative outfits. Therefore, the other compatible items may be selected as negative samples so that the diverisity decreases. However, TryonCM2 is trained only relied on the positive outfits and avoids the above issue (Table 3.2).

   Diversity Visualization. We visualized the diversity of both the compatible item retrieval and outfit recommendation tasks. The former one shows whether the retrieved items for

Table 3.2 Diversity results of TryOn-CM and our proposed TryonCM2

	ILD@k			F_m@k		
	k = 10	k = 50	k = 100	k = 10	k = 50	k = 100
TryOn-CM	0.124	0.157	0.160	0.137	0.183	0.192
TryonCM2	0.125	0.159	0.162	0.181	0.223	0.230

a query are diverse, while the latter one shows whether the outfits that gain the similar compatibility scores are diverse.

Compatible Item Retrieval. The compatible item retrieval examples of TryonCM2 can be derived in Fig. 3.6. From Fig. 3.6, we can see that the retrieved items match well with the query item(s), although they are various in colors or styles. For example, for the second example in Fig. 3.6, TryonCM2 retrieves the blue jeans (the third item) and white pants (the fifth item) for the query green blouse to pair up as a casual look, while it also retrieves the grey pants (the eighth item) and black skirt (the ninth item) to pair up as a formal style.

Outfit Recommendation. To show the retrieval diversity of the outfit recommendation, we measured the compatibility of all outfits in our testing set and clustered the outfits according to their compatibility scores. Intuitively, outfits in the same cluster have the same compatibility score. For illustration, in Fig. 3.8, we randomly visualized 10 outfits of the two clusters that have high compatibility scores, respectively. Checking the examples, we have the following observations. On the one hand, the outfits are various in the colors, number or categories of the items, although they have the same compatibility scores. For example, the left-top outfit of cluster (a) consists of an off-white jacket and a white dress, while the center-top outfit consists of a black blazer, sleeveless vest and loose white pants. On the other hand, the outfits with different compatibility scores look different. For example, most

Outfit Cluster (a) Outfit Cluster (b)

Fig. 3.8 Two clusters of outfits that have the same high compatibility scores

of the outfits in cluster (a) are in monotonous colors and look formal, while outfits in cluster (b) are colorful and in a casual style.

3.4.6 Generation Visualization and Quality Analysis

In a sense, the performance of the combined try-on appearance modeling is depended on the quality of the synthesized try-on template. Hence, to further understand the combined try-on appearance modeling, we visualized some synthesized try-on templates in Fig. 3.9. As we can seen, most of the synthesized try-on templates are meaningful as the try-on appearance of these composing fashion items. Even through images are slightly vague, the synthesized try-on templates can help the combined compatibility modeling, since they deliver the correct information regarding the shape, color and position of each fashion item. Meanwhile, the generator performs not well in some cases as shown in the failure cases, where the synthesized try-on templates suffer from the irregular shapes and the chaos of the colors of fashion items. This may be caused by the unusual colors and complicated patterns

Fig. 3.9 Visualization of the try-on templates of both the successful and failure cases. Each box contains one outfit example, where the synthesized template is shown at the left side followed by the composing fashion items

of the fashion items that hardly appear in the training set so that the generator cannot handle these complex cases.

3.5   Conclusion and Future Work

In this work, we present a try-on-enhanced fashion compatibility modeling framework, which integrates the try-on appearance modeling into the discrete item interaction modeling. Different from TryOn-CM that learns a latent feature to represent the combined compatibility, we focus on learning the contextual structure relationships in the try-on appearance image, which can capture the intrinsic fine-grained compatibility of the outfit. Notably, this extension only utilizes the well-matched outfits, omitting the unreliable negative sampling process. Extensive experiments conducted over the FOTOS dataset demonstrate the superiority of the our proposed method.

Besides, we have found that the combined compatibility can promote fashion compatibility modeling, although try-on templates sometimes miss the details. Limitations of TryonCM2 include that we utilize a naive manner to derive the local contents of the try-on template, i.e., the image stripes of the try-on template. However, it is intuitive that the side stripes bring a little contents of the try-on appearance, while the central ones bring more. In the further, we plan to devise a adaptive method to derive the local contents of the try-on appearance and leverage a spatial method to model their relationships rather than LSTMs.

References

1. Ak, K.E., Kassim, A.A., Lim, J., Tham, J.Y.: Attribute manipulation generative adversarial networks for fashion images. In: Proceedings of the IEEE/CVF International Conference on Computer Vision, pp. 10540–10549 (2019)
2. Chaidaroon, S., Fang, Y., Xie, M., Magnani, A.: Neural compatibility ranking for text-based fashion matching. In: Proceedings of the International ACM SIGIR Conference on Research and Development in Information Retrieval, pp. 1229–1232. ACM (2019)
3. Cremonesi, P., Koren, Y., Turrin, R.: Performance of recommender algorithms on top-n recommendation tasks. In: Proceedings of the ACM Conference on Recommender Systems, pp. 39–46 (2010)
4. Cui, Z., Li, Z., Wu, S., Zhang, X.Y., Wang, L.: Dressing as a whole: Outfit compatibility learning based on node-wise graph neural networks. In: Proceedings of the ACM International Conference on World Wide Web Conference, pp. 307–317 (2019)
5. Dong, M., Zhou, D., Ma, J., Zhang, H.: Towards intelligent design: A self-driven framework for collocated clothing synthesis leveraging fashion styles and textures. In: Proceedings of the IEEE International Conference on Acoustics, Speech, and Signal Processing (2024)
6. Dong, X., Song, X., Feng, F., Jing, P., Xu, X.S., Nie, L.: Personalized capsule wardrobe creation with garment and user modeling. In: Proceedings of the ACM International Conference on Multimedia, pp. 302–310. ACM (2019)

7. Dong, X., Wu, J., Song, X., Dai, H., Nie, L.: Fashion compatibility modeling through a multi-modal try-on-guided scheme. In: Proceedings of the International ACM SIGIR Conference on Research and Development in Information Retrieval, pp. 771–780 (2020)

8. E, R.D., E, H.G., J, W.R.: Learning representations by back-propagating errors. Nature **303**(6088), 533–536 (1986)

9. Feng, Z., Yu, Z., Yang, Y., Jing, Y., Jiang, J., Song, M.: Interpretable partitioned embedding for customized multi-item fashion outfit composition. In: Proceedings of the ACM International Conference on Multimedia Retrieval, pp. 143–151 (2018)

10. Goodfellow, I., Pouget-Abadie, J., Mirza, M., Xu, B., Warde-Farley, D., Ozair, S., Courville, A., Bengio, Y.: Generative adversarial nets. In: Proceedings of the Advances in Neural Information Processing Systems, vol. 27, pp. 1–9 (2014)

11. Han, X., Song, X., Yao, Y., Xu, X.S., Nie, L.: Neural compatibility modeling with probabilistic knowledge distillation. IEEE Transactions on Image Processing **29**, 871–882 (2019)

12. Han, X., Song, X., Yin, J., Wang, Y., Nie, L.: Prototype-guided attribute-wise interpretable scheme for clothing matching. In: Proceedings of the International ACM SIGIR Conference on Research and Development in Information Retrieval, pp. 785–794 (2019)

13. Han, X., Wu, Z., Huang, W., Scott, M.R., Davis, L.: Finet: Compatible and diverse fashion image inpainting. In: Proceedings of the IEEE/CVF International Conference on Computer Vision, pp. 4480–4490 (2019)

14. Han, X., Wu, Z., Jiang, Y.G., Davis, L.S.: Learning fashion compatibility with bidirectional LSTMs. In: Proceedings of the ACM International Conference on Multimedia, pp. 1078–1086 (2017)

15. Han, X., Wu, Z., Wu, Z., Yu, R., Davis, L.S.: VITON: an image-based virtual try-on network. In: Proceedings of the IEEE/CVF International Conference on Computer Vision and Pattern Recognition, pp. 7543–7552 (2018)

16. Huang, Z., Xu, X., Zhu, H., Zhou, M.: An efficient group recommendation model with multiattention-based neural networks. IEEE Transactions on Neural Networks and Learning Systems **31**(11), 4461–4474 (2020)

17. Isola, P., Zhu, J.Y., Zhou, T., Efros, A.A.: Image-to-image translation with conditional adversarial networks. In: Proceedings of the IEEE/CVF Conference on Computer Vision and Pattern Recognition, pp. 1125–1134 (2017)

18. Krizhevsky, A., Sutskever, I., Hinton, G.E.: Imagenet classification with deep convolutional neural networks. In: Proceedings of the Advances in Neural Information Processing Systems, pp. 1106–1114 (2012)

19. Li, X., Wang, X., He, X., Chen, L., Xiao, J., Chua, T.S.: Hierarchical fashion graph network for personalized outfit recommendation. In: Proceedings of the International ACM SIGIR Conference on Research and Development in Information Retrieval, pp. 159–168 (2020)

20. Lin, Y., Ren, P., Chen, Z., Ren, Z., Ma, J., de Rijke, M.: Improving outfit recommendation with co-supervision of fashion generation. In: Proceedings of the ACM International Conference on World Wide Web Conference, pp. 1095–1105 (2019)

21. Lin, Y.L., Tran, S., Davis, L.S.: Fashion outfit complementary item retrieval. In: Proceedings of the IEEE/CVF Conference on Computer Vision and Pattern Recognition, pp. 3311–3319 (2020)

22. Liu, J., Lu, H.: Deep fashion analysis with feature map upsampling and landmark-driven attention. In: Proceedings of the European Conference on Computer Vision Workshops, pp. 30–36 (2018)

23. Liu, J., Song, X., Ren, Z., Nie, L., Tu, Z., Ma, J.: Auxiliary template-enhanced generative compatibility modeling. In: Proceedings of the International Joint Conference on Artificial Intelligence, pp. 3508–3514 (2020)

24. Liu, L., Zhang, H., Xu, X., Zhang, Z., Yan, S.: Collocating clothes with generative adversarial networks cosupervised by categories and attributes: A multidiscriminator framework. IEEE Transactions on Neural Networks and Learning Systems **31**(9), 3540–3554 (2019)

25. Liu, Y., Xiao, Y., Wu, Q., Miao, C., Zhang, J., Zhao, B., Tang, H.: Diversified interactive recommendation with implicit feedback. In: Proceedings of the AAAI Conference on Artificial Intelligence, pp. 4932–4939 (2020)
26. P, K.D., M, W.: Auto-encoding variational bayes. In: Proceedings of the International Conference on Learning Representations (2014)
27. Peng, C., Gao, X., Wang, N., Tao, D., Li, X., Li, J.: Multiple representations-based face sketch-photo synthesis. IEEE Transactions on Neural Networks and Learning Systems 27(11), 2201–2215 (2016)
28. Reed, S.E., Akata, Z., Yan, X., Logeswaran, L., Schiele, B., Lee, H.: Generative adversarial text to image synthesis. In: International Conference on Machine Learning, pp. 1060–1069 (2016)
29. Ronneberger, O., Fischer, P., Brox, T.: U-net: Convolutional networks for biomedical image segmentation. In: Proceedings of the Medical Image Computing and Computer-Assisted Intervention Conference, pp. 234–241 (2015)
30. Shen, W., Liu, R.: Learning residual images for face attribute manipulation. In: Proceedings of the IEEE/CVF International Conference on Computer Vision and Pattern Recognition, pp. 1225–1233 (2017)
31. Shu, X., Zhang, L., Qi, G.J., Liu, W., Tang, J.: Spatiotemporal co-attention recurrent neural networks for human-skeleton motion prediction. IEEE Transactions on Pattern Analysis and Machine Intelligence 44(6), 3300–3315 (2021)
32. Simonyan, K., Zisserman, A.: Very deep convolutional networks for large-scale image recognition. In: Proceedings of the International Conference on Learning Representations (2015)
33. Song, X., Feng, F., Han, X., Yang, X., Liu, W., Nie, L.: Neural compatibility modeling with attentive knowledge distillation. In: Proceedings of the International ACM SIGIR Conference on Research and Development in Information Retrieval, pp. 5–14 (2018)
34. Song, X., Feng, F., Liu, J., Li, Z., Nie, L., Ma, J.: Neurostylist: Neural compatibility modeling for clothing matching. In: Proceedings of the ACM International Conference on Multimedia, pp. 753–761 (2017)
35. Vasileva, M.I., Plummer, B.A., Dusad, K., Rajpal, S., Kumar, R., Forsyth, D.: Learning type-aware embeddings for fashion compatibility. In: Proceedings of the European Conference on Computer Vision, pp. 390–405 (2018)
36. Vincent, P., Larochelle, H., Bengio, Y., Manzagol, P.: Extracting and composing robust features with denoising autoencoders. In: International Conference on Machine Learning, pp. 1096–1103 (2008)
37. Wang, W., Xu, Y., Shen, J., Zhu, S.: Attentive fashion grammar network for fashion landmark detection and clothing category classification. In: Proceedings of the IEEE/CVF International Conference on Computer Vision and Pattern Recognition, pp. 4271–4280 (2018)
38. Wang, X., Zhu, L., Yang, Y.: T2VLAD: global-local sequence alignment for text-video retrieval. In: Proceedings of the IEEE/CVF International Conference on Computer Vision and Pattern Recognition, pp. 5079–5088 (2021)
39. Wei, Y., Wang, X., Guan, W., Nie, L., Lin, Z., Chen, B.: Neural multimodal cooperative learning toward micro-video understanding. IEEE Transactions on Image Processing 29, 1–14 (2019)
40. Wei, Y., Wang, X., Nie, L., He, X., Hong, R., Chua, T.S.: Mmgcn: Multi-modal graph convolution network for personalized recommendation of micro-video. In: Proceedings of the ACM International Conference on Multimedia, pp. 1437–1445 (2019)
41. Yang, X., Du, X., Wang, M.: Learning to match on graph for fashion compatibility modeling. In: Proceedings of the AAAI Conference on Artificial Intelligence, pp. 287–294 (2020)
42. Yang, X., He, X., Wang, X., Ma, Y., Feng, F., Wang, M., Chua, T.: Interpretable fashion matching with rich attributes. In: Proceedings of the International ACM SIGIR Conference on Research and Development in Information Retrieval, pp. 775–784 (2019)

43. Yang, X., Ma, Y., Liao, L., Wang, M., Chua, T.S.: Transnfcm: Translation-based neural fashion compatibility modeling. In: Proceedings of the AAAI Conference on Artificial Intelligence, pp. 403–410 (2019)
44. Yang, X., Song, X., Han, X., Wen, H., Nie, J., Nie, L.: Generative attribute manipulation scheme for flexible fashion search. In: Proceedings of the International ACM SIGIR Conference on Research and Development in Information Retrieval, pp. 941–950. ACM (2020)
45. Zhang, M., Wang, N., Li, Y., Gao, X.: Neural probabilistic graphical model for face sketch synthesis. IEEE Transactions on Neural Networks and Learning Systems **31**(7), 2623–2637 (2020)
46. Zheng, N., Song, X., Chen, Z., Hu, L., Cao, D., Nie, L.: Virtually trying on new clothing with arbitrary poses. In: Proceedings of the ACM International Conference on Multimedia, pp. 266–274 (2019)
47. Zhou, D., Zhang, H., Ma, J., Shi, J.: Bc-gan: A generative adversarial network for synthesizing a batch of collocated clothing. IEEE Transactions on Circuits and Systems for Video Technology pp. 1–15 (2023)
48. Zhu, J., Zhang, R., Pathak, D., Darrell, T., Efros, A.A., Wang, O., Shechtman, E.: Toward multimodal image-to-image translation. In: Proceedings of the Advances in Neural Information Processing Systems, pp. 465–476 (2017)
49. Zhu, J.Y., Park, T., Isola, P., Efros, A.A.: Unpaired image-to-image translation using cycle-consistent adversarial networks. In: Proceedings of the IEEE/CVF International Conference on Computer Vision, pp. 2223–2232 (2017)

Fine-Grained Fashion Compatibility Modeling

4

4.1 Introduction

In this chapter, we explore the automatic modeling of fine-grained outfit compatibility. With the blossoming of e-commerce, a growing number of people are turning to online shopping for fashion garments. Enjoying the convenience provided by the online fashion market, people also tend to be overwhelmed by the numerous online clothing items. Specifically, they are frequently encountered with problems such as: "does this T-shirt match my jeans" or "which shirt is better for the skirt". Towards this end, fashion compatibility modeling (FCM) that aims to automatically assess the compatibility (i.e., matching score) of items in an outfit has gained growing research attention. There have been numerous works of FCM proposed in the literature [2, 4, 9, 16, 18, 20, 28, 34]. Despite the impressive progress, current FCM methods still perform far from satisfactory. They only focus on modeling the compatibility relationship among discrete items in an outfit but overlook the human habits on fashion compatibility evaluation. In fact, people usually evaluate the matching degree of a given outfit from not only the collocation angle (i.e., in a discrete manner) but also the try-on angle (i.e., in a unified manner). In light of this, we aim to devise a comprehensive FCM scheme that evaluates the outfit compatibility from both the discrete collocation and unified try-on angles, as shown in Fig. 4.1.

However, combining both the collocation and try-on angles is non-trivial due to the following challenges: (i) the visual compatibility among discrete items in an outfit can be affected by multiple latent factors (e.g., color, material and style), which indicates that the overall compatibility among items can be decoupled into multiple latent fine-grained visual compatibility. We argue that exploring the latent fine-grained compatibility would make the task more tractable, thus improving model performance. Therefore, capturing the latent fine-grained compatibility among discrete items to enhance FCM is the first challenge; (ii) a simple solution towards try-on compatibility modeling is to evaluate the compatibility through an outfit's real try-on appearance image. However, the real try-on appearance is

W. Guan et al., *Advanced Multimodal Compatibility Modeling and Recommendation*,
Synthesis Lectures on Information Concepts, Retrieval, and Services,
https://doi.org/10.1007/978-3-031-81048-0_4

Fig. 4.1 Illustration of the compatibility modeling from both collocation and try-on perspectives

usually unavailable in practice. Therefore, how to utilize the limited training try-on appearance images of outfits to model the try-on compatibility for outfits without the real try-on images constitutes another challenge; and (iii) the compatibility of the same outfit evaluated from the collocation and try-on angles should be intrinsically consistent. Therefore, utilizing the latent consistency to integrate the collocation and try-on compatibility modeling seamlessly, in order to boost the model performance, poses the third challenge.

To address the aforementioned challenges, we propose the **C**ollocation and **T**ry-**O**n **Net**work (CTO-Net) for FCM. As shown in Fig. 4.2, CTO-Net consists of two core parts: *Collocation Compatibility Modeling (CCM)* and *Try-on Compatibility Modeling (TCM)*. CCM implicitly fulfills the fine-grained compatibility modeling. Specifically, to uncover the latent factors influencing the compatibility, CCM injects disentangled item representations into a graph convolutional network (GCN) and devises a new disentangled compatibility propagation module to adaptively propagate the fine-grained compatibility relationships among items. TCM fulfills the try-on representation learning by exploiting the teacher-student knowledge distillation scheme. In particular, the teacher network is first trained using unsupervised self-encoding. Then, the student network imitates the output of the teacher network and hence derives the accurate try-on representation directly from the outfit's discrete composing items. To strengthen the try-on representation learning, we incorporate the category information of each item as the context, which remains untapped by previous studies. In addition, we employ the mutual learning strategy [42] to encourage both the CCM and TCM to transfer knowledge from each other and further boost the final compatibility modeling performance. Experimental results on the real-world dataset demonstrate the superiority of our CTO-Net over the state-of-the-art methods.

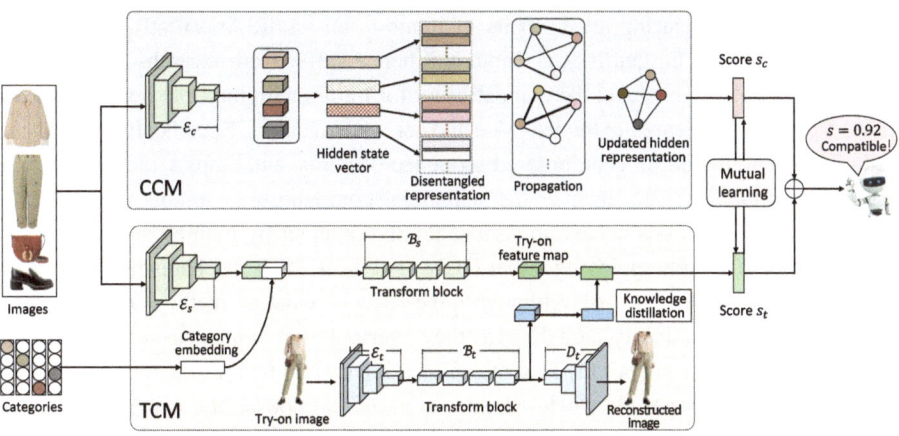

Fig. 4.2 Illustration of the proposed CTO-Net, which consists of two core parts: CCM and TCM. In particular, CCM and TCM are responsible for modeling the fine-grained collocation compatibility and the unified try-on compatibility, respectively. Finally, the mutual learning is employed to encourage the knowledge sharing between CCM and TCM

The main contributions of this research are summarized as follows:

1. Inspired by the human habits towards the fashion compatibility evaluation, we present a novel framework, i.e., CTO-Net, to comprehensively analyze the fashion compatibility from both the discrete collocation and unified try-on angles. In particular, the two compatibility modeling schemes get mutually enhanced by absorbing knowledge from each other.
2. We propose a novel disentangled graph learning scheme, which is capable of analyzing the fine-grained collocation compatibility through propagating the compatibility between discrete items based on their disentangled representations.
3. We propose an integrated distillation learning scheme for try-on compatibility modeling, which utilizes the limited try-on appearance images for guiding the network to learn a reliable try-on representation based on the discrete fashion items in the outfit.

4.2   Related Work

This work is related to fashion compatibility modeling, graph convolutional network, and knowledge distillation.

Fashion Compatibility Modeling. Existing studies on FCM can be summarized into three groups, i.e., pair-wise [22, 25, 28, 29, 31], list-wise [5, 6, 9], and set-wise [4, 16, 18]. The pair-wise methods focus on compatibility between a pair of items and derive the compatibility by measuring all the pairs of items in an outfit. Apparently, the pair-wise methods do not treat the outfit as a whole and hence suffer from the sub-optimal performance. As to directly evaluate the compatibility for outfits with multiple items, increasing efforts have been dedicated to the list-wise and set-wise manners. Specifically, the list-wise methods assume the outfit as an ordered sequence of items, and employ the bi-directional LSTM [9] or GRU [2], to uncover the sequential compatibilities among items. Beyond that, the set-wise methods behave in a more flexible manner by treating an outfit as a set of unordered items and employing either GCN or self-attention mechanism to evaluate an outfit's matching degree [4, 16]. Although huge success, existing methods cannot achieve comprehensive compatibility modeling, as they overlook that humans usually evaluate the outfit compatibility from both the discrete collocation and unified try-on angles.

Graph Convolutional Network. Mathias et al. [27] proposed the graph convolutional network (GCN), which generalizes convolution operation to the graph domain [14]. Due to its remarkable capability of representation learning, GCN has been widely explored in various tasks, including recommendation [32], information retrieval [7], and visual comprehension [36, 38]. Recently, in the fashion domain, as each outfit can be abstracted as an item graph, several GCN-based methods, like NGNN [4] and HFGN [16], have been proposed for FCM. The key of these methods is to update the item embedding with its context (the other fashion items) in the outfit. Different from these methods that only propagate the general item embedding, in this work, we conducted the fine-grained item-item relationship propagation among items with the disentangled representation learning.

Knowledge Distillation. Due to the remarkable performance in various tasks [12, 15, 33, 39], knowledge distillation has attracted growing research attention recently. Knowledge distillation adopts the teacher-student network and aims to enhance the student network by transferring the knowledge from a powerful teacher network to the student network [10]. Recently, several studies have been dedicated to exploring the potency of the knowledge distillation to enhance the performance in the fashion domain [1, 28]. For example, Song et al. [28] proposed to incorporate fashion domain knowledge to facilitate fashion compatibility evaluating. Beyond the existing efforts, we worked on transferring the try-on knowledge gained from the teacher network with the real try-on image, to improve the try-on representation directly based on the outfit's discrete composing items.

4.3 Methodology

4.3.1 Problem Formulation

Suppose that we have the training set $\Omega = \{(O^i, y^i)|i = 1, \ldots, N\}$ composed of N outfits, where O^i is the i-th outfit, and y^i denotes the ground truth label. Specifically, $y^i = 1$ indicates that the outfit O^i is compatible and $y^i = 0$ otherwise. Each outfit O^i consists of M_i complementary fashion items $O^i = \{o^i_1, o^i_2, \ldots, o^i_{M_i}\}$, where o^i_j is the j-th item, associated with an image pixel array $\mathbf{o}_j^i$, and a category embedding $\mathbf{c}_j^i$. In particular, $\mathbf{c}_j^i \in \mathbb{R}^C$ is a one-hot vector and C is the total number of fashion item categories in the dataset. Besides, only the positive/compatible outfits, have their corresponding try-on appearance images. Accordingly, the whole training set Ω can be split into two sets: one set of positive outfits $\Omega_+ = \{(O^i, \mathbf{P}^i, y^i) | y^i = 1\}$ and one set of negative outfits $\Omega_- = \{(O^i, y^i) | y^i = 0\}$. $\mathbf{P}^i$ denotes the i-th outfit's try-on image pixel array. Based on these data, we aim to devise a FCM scheme $\mathcal{F}$ to evaluate the compatibility score s^i of a given outfit $O^i = \{o_1^i, o_2^i, \ldots, o_{M_i}^i\}$ as follows:

$$s^i = \mathcal{F}(\{o_j^i\}_{j=1}^{M_i} | \Theta), \tag{4.1}$$

where Θ is a set of to-be-learned parameters. Notably, we omit the superscript i in the rest of the chapter for brevity.

4.3.2 Disentangled Graph Learning for CCM

Existing methods [16] utilize general representations of composing items to capture the underlying collocation compatibility. However, we argue that the compatibility relationship among discrete items can be influenced by multiple latent factors, like color, texture, and style. Hence, there simultaneously exist multiple latent fine-grained compatibility relationships among discrete items. In light of this, we propose the disentangled graph learning scheme for CCM, which consists of three key parts: graph initialization, disentangled item representation, and disentangled compatibility propagation.

Graph Initialization. For each outfit O, we first construct an undirected graph $\mathcal{G} = (\mathcal{H}, \mathcal{E})$, where $\mathcal{H} = \{h_j\}_{j=1}^M$ is the set of nodes corresponding to the M composing fashion items in the outfit, respectively, while $\mathcal{E} = \{(h_j, h_k) | j, k \in [1, \ldots, M], j \neq k\}$ is the set of edges indicating the relation among the composing items of the outfit. Each node h_j is associated with a hidden state vector $\mathbf{h}_j$ utilized for the compatibility propagation. Since the visual cue plays an important role in the compatibility modeling, we initialize the hidden state vector $\mathbf{h}_j$ with the visual feature of the corresponding item o_j. Specifically, we introduce a visual encoder $\mathcal{E}_c$, consisting of several convolutional layers to obtain the visual feature map $\mathbf{F}_j$ of each fashion item o_j. Then we obtain the visual representation $\mathbf{f}_j$ of the j-th item by reshaping the visual feature map $\mathbf{F}_j$ into a vector. Ultimately, we initialize the hidden state vector $\mathbf{h}_j$ with a linear transformation over the visual representation $\mathbf{f}_j$ as follows:

$$\mathbf{h}_j = \frac{\mathbf{W}_h \mathbf{f}_j + \mathbf{b}_h}{||\mathbf{W}_h \mathbf{f}_j + \mathbf{b}_h||_2}, \tag{4.2}$$

where $\mathbf{W}_h$ and $\mathbf{b}_h$ are the weight matrix and bias vector to be learned, respectively. The linear transformation aims to project the visual representation to a low-dimensional space and the normalization is used to ensure the numerical stability.

Disentangled Item Representation. We argue that utilizing a single overall hidden vector to capture the multi-facet fine-grained compatibility relationship among fashion items can be insufficient. Beyond existing studies, inspired by the recent advance of disentangled representation learning in various recommendation tasks [11, 32], we propose to disentangle the multi-facet compatibility relationship among discrete items and capture their fine-grained compatibility on the basis of GCN.

In particular, we suppose that there are L latent factors $\{f_1, f_2, \ldots, f_L\}$ influencing the compatibility relationship among items. For each latent factor f_l, we employ a condition mask $\mathbf{C}_l \in \mathbb{R}^d$ to derive the disentangled item representation $\mathbf{u}_j^l$ pertaining to the l-th latent factor as follows:

$$\mathbf{u}_j^l = \mathbf{h}_j \odot \mathbf{C}_l, \tag{4.3}$$

where $\odot$ denotes the element-wise product.

Intuitively, to promote the fine-grained compatibility modeling for CCM, we expect that each disentangled representation can focus on only one latent factor, namely, the disentangled representations for different latent factors should be as independent as possible. Therefore, we utilize L1 regularization on the condition masks to encourage the sparsity and disentanglement [30, 34] as follows:

$$\mathcal{L}_{mask} = \frac{1}{L} \sum_{l=1}^{L} ||\mathbf{C}_l||_1. \tag{4.4}$$

Disentangled Compatibility Propagation. After obtained the disentangled representation for each node (item), we proceed to the disentangled compatibility propagation over the graph. In particular, we employ L-way parallel propagation to deliver the fine-grained compatibility between items, as shown in Fig. 4.3. One naive way to fulfill the information propagation between nodes o_j and o_k is to equally aggregate all the parallel propagations. However, the factor that dominates the compatibility between different item pairs may be different. For example, as shown in Fig. 4.4, the factor that influences the compatibility for the left pair of items is most likely to be the pattern, while that for the right one is the style. In light of this, we incorporate the factor importance in the disentangled compatibility relationship propagation with the attention mechanism [21, 23, 24, 26, 35]. Specifically, given

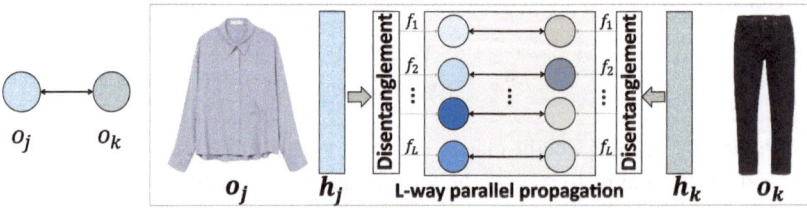

Fig. 4.3 Illustration of L-way parallel propagation between two nodes (items)

Fig. 4.4 Examples of different factors that dominate the compatibility between different items

items o_j and o_k, we assess the attention score for their collocation compatibility on the l-th latent factor as follows:

$$
\begin{cases}
\mathbf{e}_{j,k} = \mathbf{W}_a^2(\delta(\mathbf{W}_a^1(\mathbf{h}_j||\mathbf{h}_k) + \mathbf{b}_a^1)) + \mathbf{b}_a^2, \\
a_{j,k}^l = \dfrac{\exp(e_{j,k}^l)}{\sum_{p=1}^{L} \exp(e_{j,k}^p)}, l \in \{1, 2, \ldots, L\},
\end{cases}
\tag{4.5}
$$

where $||$ denotes the concatenation operator; $\mathbf{W}_a^1$, $\mathbf{W}_a^2$, $\mathbf{b}_a^1$, and $\mathbf{b}_a^2$ refer to the attention network parameters. $\delta(\cdot)$ denotes the Tanh activate function. $\mathbf{e}_{j,k} = [e_{j,k}^1, e_{j,k}^2, \ldots, e_{j,k}^L] \in \mathbb{R}^L$ is the vector of intermediate attention scores for L factors.

Based on these attention scores, we define the message passing from the item o_k to the item o_j as follows:

$$
\mathbf{m}_{k \to j} = \phi\left(\mathbf{W}_p \sum_{l=1}^{L} a_{j,k}^l(\mathbf{u}_j^l \odot \mathbf{u}_k^l) + \mathbf{b}_p\right),
\tag{4.6}
$$

where $\mathbf{W}_p$ and $\mathbf{b}_p$ denote the weight and bias to be learned, respectively. $\phi(\cdot)$ denotes the LeakyReLU activate function. Notably, instead of directly propagating the embedding of the k-th item to the j-th item, we focus more on the interaction between them, which is supposed to be the underlying compatibility between them. In particular, $\mathbf{u}_j^l \odot \mathbf{u}_k^l$ accounts for the compatibility relationship between items o_j and o_k in terms of the l-th latent factor.

Then by summarizing the passing message from all neighbors, the hidden state vector of item o_j can be updated as follows:

$$
\mathbf{h}_j^* = \phi(\mathbf{W}_e\mathbf{h}_j + \mathbf{b}_e) + \sum_{o_k \in \mathcal{N}_j} \frac{1}{|\mathcal{N}_j|} \mathbf{m}_{k \to j},
\tag{4.7}
$$

where $\mathbf{W}_e$ and $\mathbf{b}_e$ denote the weight matrix and bias to be learned, respectively. $\mathcal{N}_j$ stands for the set of neighbor nodes of the node o_j, i.e., all the complementary items for item o_j in outfit O. $\mathbf{h}_j^* \in \mathbb{R}^d$ is the updated hidden representation of item o_j, which absorbs information from both the neighbors and the node itself.

Finally, we feed the updated item representation that embodies its compatibility towards all the other items in the outfit to a multi-layer perceptron (MLP) to acquire its compatibility score towards the whole outfit. Thereafter, by summing the compatibility scores of all items in the outfit, we can derive the outfit's collocation compatibility. Formally, we have:

$$s_c = \frac{1}{M} \sum_{j=1}^{M} \sigma(\mathbf{W}_c^2(\phi(\mathbf{W}_c^1\mathbf{h}_j^*))), \tag{4.8}$$

where $\sigma(\cdot)$ denotes the Sigmoid function to normalize the compatibility score. $\mathbf{W}_c^1$ and $\mathbf{W}_c^2$ are the layer parameters.

4.3.3   Integrated Distillation Learning for TCM

The real try-on appearance image of an outfit is usually unavailable in practice. Although existing method [6] directly generates the outfit try-on appearance image based on the discrete items with a template generator, it suffers from the large input-output misalignment of the template generator. Instead, we resort to the teacher-student knowledge distillation scheme [33]. In particular, the teacher network is to learn the reliable try-on representation by unsupervised self-encoding with the target try-on appearance image, while the student network is to learn the global try-on representation with the guidance of the teacher network.

Teacher Network. To learn a valid representation of the try-on appearance and to guide the student network, we employ the auto-encoder network [40] as the teacher network, which has proven to be effective in the unsupervised visual encoding [41]. In particular, we adopt the ResNet-like architecture that has shown remarkable performance in various generation tasks [43, 44] as the teacher network $\mathcal{T}$. To be more specific, the teacher network comprises an encoder $\mathcal{E}_t$ for compressing the real try-on appearance image $\mathbf{P}$ of the outfit O into the visual feature map as $\mathbf{z}_t = \mathcal{E}_t(\mathbf{P})$, a transform block $\mathcal{B}_t$ for converting the visual feature map into the more expressive one as $\mathbf{b}_t = \mathcal{B}_t(\mathbf{z}_t)$, and a decoder $\mathcal{D}_t$ for reconstructing the original try-on appearance image based on $\mathbf{b}_t$ as $\hat{\mathbf{P}} = \mathcal{D}_t(\mathbf{b}_t)$. For optimization, we introduce L1 regularization to minimize the discrepancy between the reconstructed try-on appearance $\hat{\mathbf{P}}$ and the ground truth one $\mathbf{P}$ as follows:

$$\mathcal{L}_t = \sum_{\mathbf{P} \in \mathcal{P}_+} ||\hat{\mathbf{P}} - \mathbf{P}||_1. \tag{4.9}$$

Student Network. Beyond existing studies that only focus on the visual cues of fashion items [6], we also take into account the category metadata in the try-on appearance learning. The major concern is that the category information of fashion items plays a pivotal role in the spatial arrangement of fashion items in an outfit try-on appearance, e.g., the sweater or shirt is always on top of the trousers or jeans. Specifically, we devise the student network S with a visual encoder $\mathcal{E}_s$ for merging the visual cues of the composing items into the global visual embedding $\mathbf{z}_s$, a multi-modal fusion block $\mathcal{M}_s$ for fusing $\mathbf{z}_s$ and the category embedding into the latent feature map $\mathbf{m}_s$, and a transform block $\mathcal{B}_s$ for converting $\mathbf{m}_s$ into the try-on feature map $\mathbf{b}_s$ as follows:

$$\begin{cases} \mathbf{z}_s = \mathcal{E}_s(\mathbf{O}), \\ \mathbf{m}_s = \mathcal{M}_s(\mathbf{z}_s, \text{Rep}(\mathbf{W}_s^c \mathbf{C})), \\ \mathbf{b}_s = \mathcal{B}_s(\mathbf{m}_s), \end{cases} \tag{4.10}$$

where $\mathbf{O} = [\mathbf{o}_1, \mathbf{o}_2, \ldots, \mathbf{o}_M]$ and $\mathbf{C} = [\mathbf{c}_1, \mathbf{c}_2, \ldots, \mathbf{c}_M]$ are the visual and category cues of items in the outfit. $\mathbf{W}_s^c \in \mathbb{R}^{d \times (C \times M)}$ is the weight matrix of the linear transformation to learn the global category embedding of the outfit, while $\text{Rep}(\cdot)$ refers to replicating the global category embedding to form the category feature map with the same shape of visual feature map $\mathbf{z}_s$.

Knowledge Distillation. Regarding the knowledge distillation from teacher network to student network, it is natural to regulate the latent representation of the try-on appearance learned by both teacher and student networks to be similar. In particular, we first resort to the global average pooling (GAP) [19], which has shown remarkable performance in discriminative visual property extraction [37], to summarize the learned try-on representations of the teacher and student networks as $\mathbf{f}_s = \text{GAP}(\mathbf{b}_s)$ and $\mathbf{f}_t = \text{GAP}(\mathbf{b}_t)$, respectively. Then, using L1 regularization, we have:

$$\mathcal{L}_s = ||\mathbf{f}_s - \mathbf{f}_t||_1. \tag{4.11}$$

Similar with CCM, we employ a fully-connected layer to obtain the try-on compatibility score s_t as follows:

$$s_t = \sigma(\mathbf{W}_s \mathbf{f}_s + \mathbf{b}_s), \tag{4.12}$$

where $\mathbf{W}_s$ and $\mathbf{b}_s$ are layer parameters to be learned.

4.3.4 Mutual Learning Based Joint Optimization

Similar to [6, 17], we cast the compatibility modeling as a binary classification task. For each outfit O, we adopt the following cross-entropy losses for CCM and TCM:

$$\begin{cases} \mathcal{L}_{ce}^c = -y\log(s_c) - (1-y)\log(1-s_c), \\ \mathcal{L}_{ce}^t = -y\log(s_t) - (1-y)\log(1-s_t), \end{cases} \tag{4.13}$$

where y is the ground truth label of the outfit. $\mathcal{L}_{ce}^c$ and $\mathcal{L}_{ce}^t$ denote the classification losses for CCM and TCM, respectively.

Moreover, although CCM and TCM model the compatibility from different angles, for the same outfit, their evaluation should still be consistent. In other words, the knowledge of CCM can be used for guiding the TCM and vice versa. Therefore, we incorporate the mutual learning strategy [42] to encourage their knowledge sharing with each other. In particular, we adopt the most popular Kullback-Leibler divergence regularization as follows:

$$\begin{cases} \mathcal{L}_{kl}^{t \to c} = KL(\mathbf{p}_c || \mathbf{p}_t), \\ \mathcal{L}_{kl}^{c \to t} = KL(\mathbf{p}_t || \mathbf{p}_c), \end{cases} \tag{4.14}$$

where $\mathbf{p}_c = [s_c, 1 - s_c]^T$ and $\mathbf{p}_t = [s_t, 1 - s_t]^T$. $\mathcal{L}_{kl}^{t \to c}$ and $\mathcal{L}_{kl}^{c \to t}$ refer to the regularization for CCM and TCM, respectively.

Taking all the training samples into account, our final objective function can be formulated as follows:

$$\begin{cases} \mathcal{L}^c = \sum_{\Omega} (\mathcal{L}_{ce}^c + \mathcal{L}_{kl}^{t \to c} + \lambda_m \mathcal{L}_{mask}), \\ \mathcal{L}^t = \sum_{\Omega} (\mathcal{L}_{ce}^t + \mathcal{L}_{kl}^{c \to t}) + \sum_{\Omega_+} \mathcal{L}_s, \end{cases} \tag{4.15}$$

where λ_m refers to the trade-off hyper-parameter. Notably, $\mathcal{L}_s$ is optimized by the set of positive outfits Ω_+, since the set of negative ones is unavailable. Overall, we alternatively optimize the CCM and TCM modules. In particular, for each module optimization, only the corresponding parameters need to be optimized. Once the whole network gets well-optimized, we estimate the overall compatibility score s for a given outfit O as follows:

$$s = \frac{1}{2}(s_c + s_t). \tag{4.16}$$

4.4   Experiments

4.4.1   Experimental Settings

Dataset. For evaluation, we used the public dataset FOTOS [6], which consists of 10, 988 compatible outfits composed by 20, 318 fashion items. Each fashion item is associated with a visual image and the category metadata. Distinguished from other datasets, apart from the discrete items information, each outfit in FOTOS also has a corresponding try-on image, which enables the optimization of our try-on compatibility learning scheme. For the fair comparison, we adopted the public training/validating/testing data split provided by FOTOS. The detailed statistics are shown in Table 4.1.

Evaluation Task and Metric. Following TryOn-CM [6], we evaluated the performance of our model with two tasks: the top-n recommendation [3] and the positive/negative outfit classification, where we followed the same data settings in [6]. For evaluation metrics, we utilized the Mean Reciprocal Ranking (MRR) and the Hit Rate (HR) @1, 10, 100, and 200 for the former task, during the Area Under Curve (AUC) for the latter one.

Implementation Details. As for CCM, we devised the visual encoder $\mathcal{E}_c$ with one 1-strided convolutional layer and four 2-strided convolutional layers, where the numbers of filters are 32, 64, 128, 256, and 512, respectively. The shape of the visual feature map

Table 4.1 Data split provided by FOTOS. In FOTOS, each outfit contains 4 clothing items at most

# Item	Training	Validating	Testing	Total
2	4,674	47	486	5,207
3	4,797	50	469	5,316
4	418	3	44	465
Total	9,889	100	999	10,988

generated by $\mathcal{E}_c$ is $512 \times 8 \times 8$, and the dimension of the hidden state vector is set to 512. The number of latent factors is set to 5.

Regarding TCM, we implemented the teacher network $\mathcal{T}$ with an encoder $\mathcal{E}_t$ sharing the same network architecture with $\mathcal{E}_c$, followed by the transform block $\mathcal{B}_t$ composed of 6 residual blocks, as well as the decoder $\mathcal{D}_t$ with four 2-strided deconvolutional layers and one 1-strided convolutional layer. The numbers of filters are set to 32, 64, 128, 256, 512, 512, 512, 512, 512, 512, 512, 256, 128, 64, 32 and 3, respectively. All convolutional layers above are followed by the Instance Normalization and ReLU function, except for the last layer, which takes the Tanh function. In addition, we implemented the student network $\mathcal{S}$ with an encoder $\mathcal{E}_s$ sharing the same network architecture with $\mathcal{E}_t$, a multi-modal fusion block $\mathcal{M}_s$ by a 1-strided convolutional layer, and a transform block $\mathcal{B}_s$ by four residual blocks. The numbers of filters are set to 32, 64, 128, 256, 512, 512, 512, 512, 512 and 512, respectively. Ultimately, we obtained $\mathbf{f}_t \in \mathbb{R}^{512}$ and $\mathbf{f}_s \in \mathbb{R}^{512}$ with the GAP layer. Similar to the study [33], we first pre-trained the teacher network and then fixed the teacher network for guiding the student network learning.

For optimization, we adopted the Adam [13] optimizer with $\beta_1 = 0.5$, $\beta_2 = 0.999$, a fixed learning rate of 0.0002, and the batch size of 32 for all experiments. As for outfits with less than 4 items, we used zero-paddings. Ultimately, we empirically found that the proposed method achieves the optimal performance with the hyper-parameter λ_m in Eq. (15) as 0.0005. In particular, we reported the average results of eight dependent experiments of our method.

4.4.2 On Model Comparison

To verify the effectiveness of our CTO-Net, we adopted the following state-of-the-art methods for comparison.

BRR-DAE [29] aims to model the coherent relation among multi-modalities of items based on a dual auto-encoder network.

PAICM [8], as a pair-wise method, utilizes matrix factorization to learn several compatible/incompatible prototypes to promote the explainable fashion compatibility modeling.

CSA-Net [30] introduces a category-based subspace attention network to flexibly learn latent representations for pair-wise fashion items based on the category labels.

NCR [2] explores the textual information in terms of the semantic and lexical aspects towards outfit compatibility modeling, where the outfit compatibility is learned in a list-wise manner.

LSTM-VSE [9] exploits the latent discrete item interaction by a bi-directional LSTM and visual-semantic consistency to facilitate the outfit compatibility modeling.

TryOn-CM [6] is the first to leverage the try-on appearance image of an outfit to boost the performance of fashion compatibility.

NGNN [4] is the first attempt to employ a graph to uncover the complex relationships among multiple complementary items.

HFGN [16] develops a hierarchical fashion graph network to unify the fashion compatibility modeling and personalized outfit recommendation. In our context, we only employed the item graph module for compatibility estimation.

CANN [18] learns the computational visual coherence by fully exploring the attention mechanism in a set-wise manner.

Table 4.2 shows the performance comparison among different methods, where the statistically significant test is performed between CTO-Net with the strongest baselines (highlighted with the underline). It is worth noting that since BPR-DAE, PAICM, NCR, LSTM-VSE, and TryOn-CM have been also adopted by the work [6], we directly referred to their performance in [6]. Note that we employed the same data settings with [6] for all experiments. From Table 4.2, we have the following observations: (1) Our CTO-Net consistently outperforms all the baselines, including both pair-wise, list-wise, and set-wise methods, by a

Table 4.2 Performance comparison among different methods

Method	AUC	MRR	HR			
			@1	@10	@100	@200
BRR-DAE	0.742	0.087	0.046	0.165	0.552	0.741
PAICM	0.692	0.057	0.024	0.110	0.468	0.662
CSA-Net	0.701	0.061	0.040	0.107	0.486	0.689
NCR	0.646	0.034	0.012	0.064	0.376	0.616
LSTM-VSE	0.794	0.118	0.065	0.226	0.642	0.809
TryOn-CM	$\underline{0.832}$	0.134	0.061	0.290	$\underline{0.721}$	$\underline{0.852}$
NGNN	0.698	0.055	0.022	0.102	0.478	0.687
HFGN	0.816	$\underline{0.154}$	0.080	$\underline{0.312}$	0.704	0.834
CANN	0.820	0.146	$\underline{0.081}$	0.267	0.702	0.837
CTO-Net	$\mathbf{0.878}^{a}$	$\mathbf{0.218}^{a}$	$\mathbf{0.134}^{a}$	$\mathbf{0.395}^{a}$	$\mathbf{0.800}^{a}$	$\mathbf{0.899}^{a}$

[a] denotes the statistical significance for $p < 0.01$, compared with the strongest baselines highlighted with the underline

large margin with respect to all metrics, which demonstrates the superiority of our proposed framework. (2) CTO-Net shows superiority over TryOn-CM, which also considers the try-on appearance to boost the compatibility modeling performance. This indicates the robustness of our integrated distillation learning scheme in obtaining the reliable try-on representation. The detailed comparison between CTO-Net and TryOn-CM will be given in Sect. 4.4.4. (3) On average, pair-wise methods (i.e., BRP-DAE, PAICM, and CSA-Net) perform worse than list-wise and set-wise methods (i.e., LSTM-VSE, TryOn-CM, HFGN, CANN, and ours). The philosophy behind may be that separately modeling the compatibility between item pairs in the outfit cannot accurately discover the complicated relationships among them and hence yields sub-optimal performance. And (4) it is unexpected that the set-wise method NGNN performs worse than the pair-wise methods. The possible reason is that NGNN focuses on propagating category-oriented fashion compatibility. Thus, it performs unsatisfactorily in our context, where the negative outfit shares the same item category as the positive one.

To gain deeper insights, we looked into the performance of our model regarding outfits with different number (i.e., M) of composing items. Figure 4.5 shows the performance comparison between our CTO-Net and the five strongest baselines, i.e., BPR-DAE, LSTM-VSE, TryOn-CM, HFGN, and CANN, with different testing configurations. As can be seen, our method surpasses all baseline methods in all settings, verifying the effectiveness of our method to handle the outfit compatibility with different composing item numbers. We also found that TryOn-CM outperforms other baselines, confirming that the try-on appearance modeling indeed benefits the FCM.

On Time Consumption. To study the efficiency of our CTO-Net, we compared the time consumption of TryOn-CM, HFGN, CANN and CTO-Net in Table 4.3. Notably, the input of all methods are images of an outfit's composing items. As can be seen, our CTO-Net shows the acceptable complexity during the training phase, while outperforms the baselines in terms of the testing phase. This demonstrates that our CTO-Net is efficient and practical for the real-world application scenarios.

Fig. 4.5 Performance comparison on compatibility evaluation in terms of different item numbers

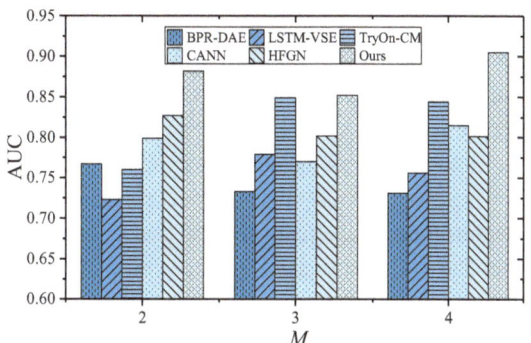

Table 4.3 The comparison of time consumption. Training(s) and Testing(s) denote the time cost for training per epoch and testing per outfit, respectively

Method	Training(s)	Testing(s)
TryOn-CM	430	0.0036
HFGN	**139**	0.0068
CANN	175	0.0079
CTO-Net	164	**0.0020**

4.4.3 On Ablation Study

To get a thorough understanding of our model, we conducted ablation experiments on the following derivatives.

CCM: This is a variant of our model that only uses the collocation compatibility modeling with the disentangled graph learning.

TCM: This is an implementation that only employs the integrated distillation learning-based TCM module.

TCM-w/o-SC and **w/o-Sc**: We removed the category input from the student network of both TCM and CTO-Net to learn its importance to try-on representation learning.

CCM-w/o-Dis and **w/o-Dis**: To validate the necessity of the fine-grained compatibility learning, we disabled the disentangled representation by setting $L = 1$ for CCM and CTO-Net, respectively.

CCM-w/o-InterP and **w/o-InterP**: To validate the function of the item-item compatibility propagation, we replaced $\mathbf{u}_j^l \odot \mathbf{u}_k^l$ with $\mathbf{u}_k^l$ in Eq.(4.6) from both CCM and CTO-Net.

CCM-w/o-Att and **w/o-Att**: To study the effect of attention mechanism in the adaptive importance attribution for different latent factors, we equally aggregated all the parallel propagations regarding different latent factors in both CCM and CTO-Net.

w/o-Mut: We removed both $\mathcal{L}_{kl}^{c \to t}$ and $\mathcal{L}_{kl}^{t \to c}$ from Eq. (4.15) to explore the importance of mutual learning strategy.

CCM-w/-Mut and **TCM-w/-Mut**: To learn the effect of mutual learning for CCM and TCM, we provided their corresponding results with the enhancement of the mutual learning, respectively.

Table 4.4 shows the ablation experimental results. Based on Table 4.4, we have the following observations. (1) Our model consistently surpasses all derivations across all metrics, demonstrating the effectiveness of each component in our proposed CTO-Net. (2) Both CTO-Net and w/o-Mut consistently surpass CCM and TCM, which implies that only modeling the compatibility from one angle (either the collocation and try-on) cannot comprehensively capture the complex compatibility relationship among multiple items. (3) CTO-Net (TCM) shows superiority over w/o-Sc (TCM-w/o-Sc). This indicates that leveraging the category information may assist to automatically capture the item spatial arrangement in

Table 4.4 The ablation experiments of our proposed method

Method	AUC	MRR	HR			
			@1	@10	@100	@200
CCM-w/o-Dis	0.825	0.125	0.063	0.241	0.719	0.847
CCM-w/o-InterP	0.786	0.076	0.031	0.154	0.638	0.805
CCM-w/o-Att	0.835	0.133	0.064	0.270	0.726	0.862
CCM	**0.848**	**0.151**	**0.079**	**0.309**	**0.751**	**0.871**
TCM-w/o-Sc	0.813	0.122	0.065	0.222	0.689	0.840
TCM	**0.835**	**0.134**	**0.072**	**0.255**	**0.723**	**0.845**
w/o-Sc	0.862	0.181	0.103	0.346	0.776	0.885
w/o-Dis	0.867	0.196	0.115	0.365	0.782	0.884
w/o-InterP	0.822	0.150	0.080	0.278	0.693	0.832
w/o-Att	0.862	0.186	0.110	0.339	0.771	0.889
CCM-w/-Mut	0.872	0.177	0.096	0.345	0.791	0.899
TCM-w/-Mut	0.854	0.175	0.099	0.319	0.764	0.873
w/o-Mut	0.867	0.190	0.112	0.346	0.784	0.893
CTO-Net	**0.878**	**0.218**	**0.134**	**0.395**	**0.800**	**0.899**

the try-on appearance, and thus boost the performance of our try-on representation learning. (4) CTO-Net (CCM) is superior to w/o-Dis (CCM-w/o-Dis), implying the necessity to explore the fine-grained compatibility between items in the FCM. (5) CTO-Net and CCM significantly outperform w/o-InterP and CCM-w/o-InterP, respectively. This confirms the facility of regarding the compatible information as a passing message rather than the item embedding in the context of outfit compatibility modeling. (6) w/o-Att (CCM-w/o-Att) is inferior to CTO-Net (CCM). The possible reason is that the confidence of the compatibility corresponding to different factors are indeed not the same, while adopting the attention mechanism can flexibly assign the factor importance. And (7) without the mutual learning strategy, w/o-Mut shows inferiority to CTO-Net, which implies that encouraging the two key modules to learn from each other can boost the overall performance of the compatibility evaluation. Meanwhile, we observed that CCM-w/-Mut and TCM-w/-Mut outperform CCM and TCM, respectively. This suggests that both CCM and TCM benefit from mutual learning.

Impact of Latent Factor Number. To study the influence of L, we conducted experiments by ranging L from 1 to 8. As shown in Fig. 4.7, the performance grows with L increasing from 1 to 5. The possible reason is that exploring the fine-grained compatibility relationships between items is beneficial to CCM. Nevertheless, the performance drops when L varies from 5 to 8, suggesting that considering too many fine-grained factors may limit the discriminative capability of the disentangled item representation.

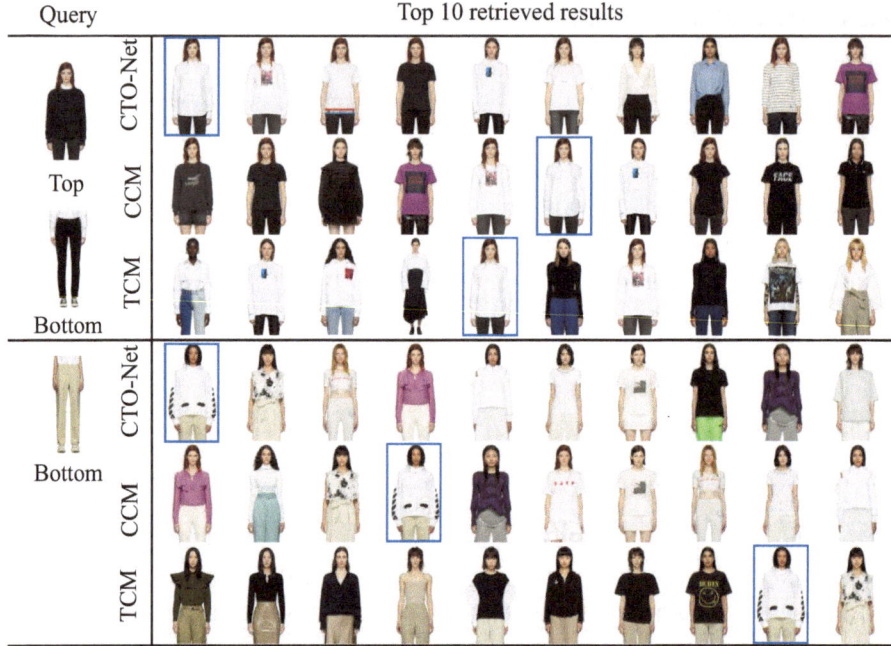

Fig. 4.6 Top-10 retrieval results of different methods

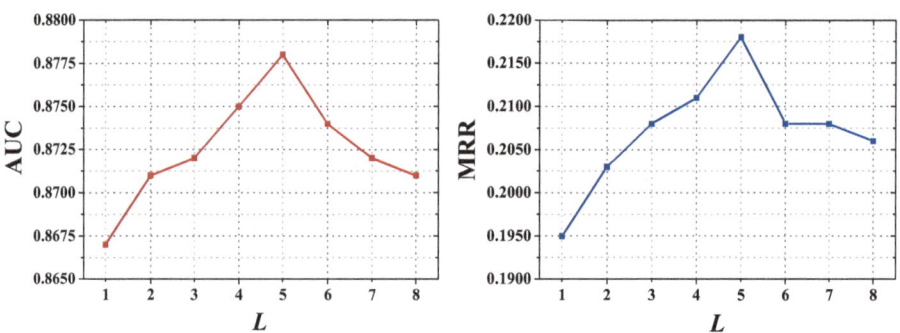

Fig. 4.7 Impact of the latent factor number L

Case Study. To get an intuitive understanding on how our model works, we compared the top-10 retrieved results of CCM, TCM, and CTO-Net for two testing queries in Fig. 4.6. The ground truth items are highlighted in the blue boxes. As can be seen, all methods can retrieve the ground truth items in a relative top ranking, indicating that both the outfit collocation and try-on reveal important cues towards FCM. In particular, CTO-Net ranks higher than CCM and TCM, respectively, implying that only exploring either perspective is insufficient for a comprehensive FCM.

Table 4.5 Performance on try-on learning methods

Method	AUC	MRR	HR			
			@1	@10	@100	@200
CCM	0.848	0.151	0.079	0.309	0.751	0.871
CCM-TG	0.854	0.173	0.103	0.323	0.760	0.874
CCM-TG-w/-Sc	0.864	0.191	0.116	0.346	0.772	0.891
CTO-Net	**0.878**	**0.218**	**0.134**	**0.395**	**0.800**	**0.899**

4.4.4 On Try-On Knowledge Distillation Study

To investigate whether the teacher-student knowledge distillation-based TCM shows superiority over the existing method, i.e., [6], we introduced the following two variants based on CTO-Net and TryOn-CM. (1) CCM-TG. In this method, we replaced the try-on knowledge distillation module of CTO-Net with the counterpart in TryOn-CM, which essentially is a try-on template generator working on producing the try-on appearance image based on the discrete composing items of an outfit. And (2) CCM-TG-w/-Sc. Since the original TryOn-CM did not incorporate the item category context in the outfit try-on looking generation, for fair comparison, we additionally fed the category context to CCM-TG in the same manner as TCM. Table 4.5 shows the performance comparison of different methods. Firstly, as can be seen, with the enhancement of the try-on modeling, both CCM-TG, CCM-TG-w/-Sc, and CTO-Net outperform CCM. This reconfirms the importance of assessing the try-on compatibility of an outfit. Secondly, we noticed that CTO-Net surpasses CCM-TG-w/-Sc, which indicates that the try-on appearance representation learned by our scheme is more reliable than that derived by a try-on template generator. Last but not least, CCM-TG achieves an inferior performance than CCM-TG-w/-Sc, validating the necessity of utilizing the item category to enhance the try-on appearance representation learning.

4.5 Conclusion and Future Work

In this work, towards outfit compatibility modeling, we present a novel collocation and try-on network, named CTO-Net, which consists of two key components: *CCM* and *TCM*. Particularly, as for the CCM, we inject the disentangled item representations into GCN and devise a novel disentangled compatibility propagation to uncover the fine-grained compatibility relationships in terms of various latent factors. Pertaining to the TCM, we introduce a teacher network to learn a real try-on representation by unsupervised self-encoding, and a student network to imitate the teacher to acquire an accurate try-on representation directly based on composing discrete items, where item category is first studied as the spatial guidance to strengthen the try-on representation learning. Furthermore, we introduce mutual learning to encourage the key two modules to transfer knowledge to each other. Extensive experiments conducted on the real-world dataset demonstrate the superiority of CTO-Net. In addition, we found that propagating the fine-grained relationships between items over the graph does greatly improve the FCM. Meanwhile, employing a teacher-student scheme for try-on representation learning is superior to existing studies. Moreover, transferring knowledge between the collocation and try-on compatibility modules is also helpful to boost the model performance. Currently, we only focus on the visual modality, but overlooking the textual context of each item. In the future, we plan to involve external information of items, such as text descriptions, to enhance the modal performance as well as interpretability.

References

1. Alashkar, T., Jiang, S., Wang, S., Fu, Y.: Examples-rules guided deep neural network for makeup recommendation. In: Proceedings of the AAAI Conference on Artificial Intelligence, pp. 941–947. AAAI (2017)
2. Chaidaroon, S., Fang, Y., Xie, M., Magnani, A.: Neural compatibility ranking for text-based fashion matching. In: Proceedings of the International ACM SIGIR Conference on Research and Development in Information Retrieval, pp. 1229–1232. ACM (2019)
3. Cremonesi, P., Koren, Y., Turrin, R.: Performance of recommender algorithms on top-n recommendation tasks. In: Proceedings of the ACM Conference on Recommender Systems, pp. 39–46 (2010)
4. Cui, Z., Li, Z., Wu, S., Zhang, X.Y., Wang, L.: Dressing as a whole: Outfit compatibility learning based on node-wise graph neural networks. In: Proceedings of the ACM International Conference on World Wide Web Conference, pp. 307–317 (2019)
5. Dong, X., Song, X., Feng, F., Jing, P., Xu, X.S., Nie, L.: Personalized capsule wardrobe creation with garment and user modeling. In: Proceedings of the ACM International Conference on Multimedia, pp. 302–310. ACM (2019)
6. Dong, X., Wu, J., Song, X., Dai, H., Nie, L.: Fashion compatibility modeling through a multi-modal try-on-guided scheme. In: Proceedings of the International ACM SIGIR Conference on Research and Development in Information Retrieval, pp. 771–780 (2020)
7. Gao, Z., Li, Y.m., Guan, W.l., Nie, W.z., Cheng, Z.y., Liu, A.a.: Pairwise view weighted graph network for view-based 3d model retrieval. In: Proceedings of the International ACM SIGIR Conference on Research and Development in Information Retrieval, pp. 129–138. ACM (2020)
8. Han, X., Song, X., Yin, J., Wang, Y., Nie, L.: Prototype-guided attribute-wise interpretable scheme for clothing matching. In: Proceedings of the International ACM SIGIR Conference on Research and Development in Information Retrieval, pp. 785–794 (2019)
9. Han, X., Wu, Z., Jiang, Y.G., Davis, L.S.: Learning fashion compatibility with bidirectional LSTMs. In: Proceedings of the ACM International Conference on Multimedia, pp. 1078–1086 (2017)
10. Hinton, G., Vinyals, O., Dean, J.: Distilling the knowledge in a neural network. arXiv preprint arXiv:1503.02531 (2015)
11. Hu, L., Xu, S., Li, C., Yang, C., Shi, C., Duan, N., Xie, X., Zhou, M.: Graph neural news recommendation with unsupervised preference disentanglement. In: Proceedings of the Annual Meeting of the Association for Computational Linguistics, pp. 4255–4264. ACL (2020)

12. Hu, Z., Ma, X., Liu, Z., Hovy H., E., Xing, E.: Harnessing deep neural networks with logic rules. In: Proceedings of the Annual Meeting of the Association for Computational Linguistics, pp. 2410–2420. ACL (2016)

13. Kingma, D.P., Ba, J.: Adam: A method for stochastic optimization. In: Proceedings of the International Conference on Learning Representations, pp. 1–15 (2014)

14. Kipf, T.N., Welling, M.: Semi-supervised classification with graph convolutional networks. In: Proceedings of the International Conference on Learning Representations. ICLR (2017)

15. Li, W.H., Bilen, H.: Knowledge distillation for multi-task learning. In: Proceedings of the European Conference on Computer Vision, pp. 163–176. Springer (2020)

16. Li, X., Wang, X., He, X., Chen, L., Xiao, J., Chua, T.S.: Hierarchical fashion graph network for personalized outfit recommendation. In: Proceedings of the International ACM SIGIR Conference on Research and Development in Information Retrieval, pp. 159–168 (2020)

17. Li, Y., Cao, L., Zhu, J., Luo, J.: Mining fashion outfit composition using an end-to-end deep learning approach on set data. IEEE Transactions on Multimedia **19**(8), 1946–1955 (2017)

18. Li, Z., Wu, B., Liu, Q., Wu, L., Zhao, H., Mei, T.: Learning the compositional visual coherence for complementary recommendations. In: Proceedings of the International Joint Conference on Artificial Intelligence, pp. 3536–3543. IJCAI (2020)

19. Lin, M., Chen, Q., Yan, S.: Network in network. arXiv:1312.4400 (2013)

20. Lin, Y.L., Tran, S., Davis, L.S.: Fashion outfit complementary item retrieval. In: Proceedings of the IEEE/CVF Conference on Computer Vision and Pattern Recognition, pp. 3311–3319 (2020)

21. Liu, F., Cheng, Z., Sun, C., Wang, Y., Nie, L., Kankanhalli, M.: User diverse preference modeling by multimodal attentive metric learning. In: Proceedings of the ACM International Conference on Multimedia, pp. 1526–1534. ACM (2019)

22. Liu, J., Song, X., Ren, Z., Nie, L., Tu, Z., Ma, J.: Auxiliary template-enhanced generative compatibility modeling. In: Proceedings of the International Joint Conference on Artificial Intelligence, pp. 3508–3514 (2020)

23. Liu, M., Nie, L., Wang, X., Tian, Q., Chen, B.: Online data organizer: Micro-video categorization by structure-guided multimodal dictionary learning. IEEE Transactions on Image Processing **28**(3), 1235–1247 (2019)

24. Liu, M., Wang, X., Nie, L., Tian, Q., Chen, B., Chua, T.S.: Cross-modal moment localization in videos. In: Proceedings of the ACM International Conference on Multimedia, pp. 843–851. ACM (2018)

25. McAuley, J., Targett, C., Shi, Q., Van Den Hengel, A.: Image-based recommendations on styles and substitutes. In: Proceedings of the International ACM SIGIR Conference on Research and Development in Information Retrieval, pp. 43–52 (2015)

26. Nie, L., Li, Y., Feng, F., Song, X., Wang, M., Wang, Y.: Large-scale question tagging via joint question-topic embedding learning. ACM Transactions on Information Systems **38**(2) (2020)

27. Niepert, M., Ahmed, M., Kutzkov, K.: Learning convolutional neural networks for graphs. In: Proceedings of the International Conference on Machine Learning, pp. 2014–2023. ACM (2016)

28. Song, X., Feng, F., Han, X., Yang, X., Liu, W., Nie, L.: Neural compatibility modeling with attentive knowledge distillation. In: Proceedings of the International ACM SIGIR Conference on Research and Development in Information Retrieval, pp. 5–14 (2018)

29. Song, X., Feng, F., Liu, J., Li, Z., Nie, L., Ma, J.: Neurostylist: Neural compatibility modeling for clothing matching. In: Proceedings of the ACM International Conference on Multimedia, pp. 753–761 (2017)

30. Tan, R., Vasileva, M.I., Saenko, K., Plummer, B.A.: Learning similarity conditions without explicit supervision. In: Proceedings of the IEEE/CVF International Conference on Computer Vision, pp. 10373–10382 (2019)

31. Vasileva, M.I., Plummer, B.A., Dusad, K., Rajpal, S., Kumar, R., Forsyth, D.: Learning type-aware embeddings for fashion compatibility. In: Proceedings of the European Conference on Computer Vision, pp. 390–405 (2018)

32. Wang, X., He, X., Wang, M., Feng, F., Chua, T.S.: Neural graph collaborative filtering. In: Proceedings of the International ACM SIGIR Conference on Research and Development in Information Retrieval, pp. 165–174. ACM (2019)

33. Wang, X., Hu, J.F., Lai, J.H., Zhang, J., Zheng, W.S.: Progressive teacher-student learning for early action prediction. In: Proceedings of the IEEE/CVF International Conference on Computer Vision and Pattern Recognition, pp. 3556–3565. IEEE (2019)

34. Wang, X., Wu, B., Zhong, Y.: Outfit compatibility prediction and diagnosis with multi-layered comparison network. In: Proceedings of the ACM International Conference on Multimedia, pp. 329–337 (2019)

35. Wei, Y., Wang, X., Guan, W., Nie, L., Lin, Z., Chen, B.: Neural multimodal cooperative learning toward micro-video understanding. IEEE Transactions on Image Processing **29**, 1–14 (2019)

36. Wei, Y., Wang, X., Nie, L., He, X., Hong, R., Chua, T.S.: Mmgcn: Multi-modal graph convolution network for personalized recommendation of micro-video. In: Proceedings of the ACM International Conference on Multimedia, pp. 1437–1445 (2019)

37. Yang, X., Song, X., Han, X., Wen, H., Nie, J., Nie, L.: Generative attribute manipulation scheme for flexible fashion search. In: Proceedings of the International ACM SIGIR Conference on Research and Development in Information Retrieval, pp. 941–950. ACM (2020)

38. Yu, J., Zhou, H., Zhan, Y., Tao, D.: Deep graph-neighbor coherence preserving network for unsupervised cross-modal hashing. In: Proceedings of the AAAI Conference on Artificial Intelligence, pp. 4626–4634. AAAI (2021)

39. Yuan, M., Peng, Y.: Ckd: Cross-task knowledge distillation for text-to-image synthesis. IEEE Transactions on Multimedia **22**(8), 1955–1968 (2019)

40. Zhan, Y., Yu, J., Yu, Z., Zhang, R., Tao, D., Tian, Q.: Comprehensive distance-preserving autoencoders for cross-modal retrieval. In: Proceedings of the ACM international conference on Multimedia, pp. 1137–1145. ACM (2018)

41. Zhang, L., Qi, G.J., Wang, L., Luo, J.: Aet vs. aed: Unsupervised representation learning by auto-encoding transformations rather than data. In: Proceedings of the IEEE/CVF International Conference on Computer Vision and Pattern Recognition, pp. 2547–2555. IEEE (2019)

42. Zhang, Y., Xiang, T., Hospedales, T.M., Lu, H.: Deep mutual learning. In: Proceedings of the IEEE/CVF International Conference on Computer Vision and Pattern Recognition, pp. 4320–4328. IEEE (2018)

43. Zheng, N., Song, X., Chen, Z., Hu, L., Cao, D., Nie, L.: Virtually trying on new clothing with arbitrary poses. In: Proceedings of the ACM International Conference on Multimedia, pp. 266–274 (2019)

44. Zhu, J.Y., Park, T., Isola, P., Efros, A.A.: Unpaired image-to-image translation using cycle-consistent adversarial networks. In: Proceedings of the IEEE/CVF International Conference on Computer Vision, pp. 2223–2232 (2017)

Attribute-Enhanced Fashion Item Recommendation

5

5.1 Introduction

In this chapter, we explore the recommendation of compatible fashion items using an attribute-enhanced schema. In recent years, the online fashion industry has flourished, with unprecedented growth in fashion data on e-commerce platforms such as Amazon and Taobao. Surfing in the huge online fashion market, people always get overwhelmed and feel difficult to find their desired items. Therefore, personalized fashion item recommendation has become an emerging need, which can not only increase the platform's economic benefits but also improve the user's consumption experience. Different from the traditional recommendation tasks, one distinct feature of the fashion item recommendation task is that it should take into account not only the user-item historical interactions but also the visual content of fashion items. This is due to the fact that the visual property of the item is an essential factor affecting the user's preference to the item. In light of this, several visual-enhanced fashion item recommendation methods have been proposed [16, 19, 39]. Although these efforts have achieved compelling success, most of them overlook the side information of items, i.e., attributes, like "color" and "category". In fact, the attributes of an item usually convey its key features, and some attributes can be even hard to be expressed by the item's visual content, like the "material" and "brand". Therefore, incorporating the attribute information to boost the item representation learning and benefit the user preference modeling merits our special attention. Meanwhile, the intuitive semantic delivered by the attribute information can be used for improving the model interpretability and facilitating users to understand the recommendation results better. Therefore, to bridge the research gap, in this work, apart from the user-item interaction histories, we also jointly consider the attribute and vision modalities of items to boost the performance and interpretability of fashion item recommendation systems.

W. Guan et al., *Advanced Multimodal Compatibility Modeling and Recommendation*, Synthesis Lectures on Information Concepts, Retrieval, and Services, https://doi.org/10.1007/978-3-031-81048-0_5

However, fulfilling the multi-modal enhanced fashion item recommendation is non-trivial due to the following three key challenges. (1) **C1: Latent Visual-Semantic Consistency.** Although the visual image and semantic attributes characterize the same fashion item from different perspectives, they should share certain latent consistency in terms of delivering the item's key features. Therefore, the primary challenge is how to effectively utilize the heterogeneous multi-modal data (i.e., the image and attributes) of items with the visual-semantic consistency modeling to enhance the user and item representation learning and thus promote the performance of fashion item recommendation. (2) **C2: Various Relation Types.** The relation types among our entities are various, including both the user-item relation (i.e., *interact*) and item-attribute relations (e.g., *hasColorOf*, *hasBrandOf*, and *hasCategoryOf*). Moreover, different relation types indicate different correlations among entities. Intuitively, different types of attributes contribute differently in characterizing the item and conveying the user's preference. Therefore, how to adaptively measure the correlation among entities taking into account their relation types poses a key challenge. (3) **C3: Unique Attributes with Insufficient Samples.** In real-world applications, not all the attributes are so common that being possessed by a large number of items. For example, only certain shoes items have the "heel" attribute. Consequently, how to properly deal with these unique attributes that have insufficient samples constitutes another challenge.

To address the aforementioned challenges, as shown in Fig. 5.1, we propose a Multi-Modal enhanced Fashion Item Recommendation scheme, termed as MM-FRec. The heterogeneous multi-modal data of items are engaged to learn user and item embeddings, and we utilize the Bayesian personalized ranking (BPR) [29] loss to fulfil the user's preference learning. In particular, for the challenge **C1**, we devise a relation-oriented user-item-attribute tripartite graph and a vision-oriented user-image bipartite graph to learn the user and item

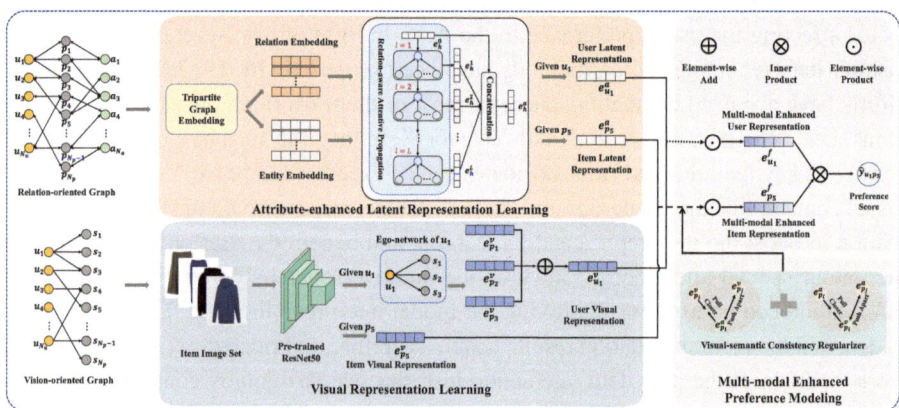

Fig. 5.1 Illustration of the proposed MM-FRec, which consists of three components: attribute-enhanced latent representation learning, visual representation learning, and multi-modal enhanced preference modeling

embeddings from the attribute and image perspectives, respectively. We enforce the embeddings of the same item from different perspectives to be similar with a visual-semantic consistency regularizer, for the purpose of enabling the learned knowledge from the two graphs to be shared. For the challenge **C2**, we design a new relation-aware attentive propagation method to learn the user and item embeddings in the user-item-attribute tripartite graph, which employs neural networks to learn the aggregating attention of neighbors taking not only the information of head and tail nodes, but the extra information of relation types as the inputs. Towards the challenge **C3**, we treat the aggregating attention weight learning of different relation types as multiple correlated tasks and design a deep multi-task learning scheme to learn the aggregating attention weights. It consists a shared network that captures the common knowledge in different relation types, and multiple relation-specific networks that learn the specific knowledge in each relation type.

Our main contributions can be summarized in threefold:

1. We present a novel Multi-Modal enhanced Fashion Item Recommendation scheme (MM-FRec) that jointly investigates the attribute-enhanced latent preference and visual preference of a user, where a visual-semantic consistency regularizer is introduced to mutually enhance the two preference learning.
2. We design a new relation-aware propagation method for the user preference learning based on GCN, where the importance of different connections during the propagation is flexibly measured by a deep multi-task learning framework and can be used for explaining the recommendation results. As far as we know, we are the first to employ the deep multi-task learning framework to enhance the connection confidence assignment for various attributes.
3. We conduct extensive experiments on a real-world dataset, and the results show that our MM-FRec achieves a prominent improvement of 28.40%, 20.22%, 24.82% in Precision@20, Precision@50, and NDCG@100, over the best baseline, respectively. As a byproduct, we have released the codes and involved parameters to benefit the community.[1]

5.2   Related Work

In this section, we briefly present the related work, including fashion item recommendation and graph-based recommendation in Sects. 5.2.1 and 5.2.2, respectively.

---

[1] https://anonymousMM-FRec.wixsite.com/MM-FRec.

5.2.1   Fashion Item Recommendation

In recent years, fashion item recommendation has been attracting increasing attention, owing to its huge economic value. Due to the fact that the visual features of items bring much positive force to reveal user preferences, many researchers have incorporated the visual images of items to boost the fashion item recommendation performance. For example, McAuley et al. [26] proposed a linear model to learn the implicit relationships between items based on their visual images. To alleviate the cold start issues, apart from the visual feature, He et al. [16] also explored the user implicit feedback (i.e., browsing logs and the temporal information) to predict the user fashion taste. Besides, Yu et al. [39] designed a brain-inspired deep network (BDN) to enhance the user preference modeling from the aesthetic perspective. Considering the semantic information of items that can intuitively express the key features of items, there are several researchers perform the multi-modal recommendation by leveraging the item images and user reviews to items [9, 10, 37]. For example, Chen et al. [9] learned an attention model over the image of fashion items and introduced user review information as a weak supervision signal to collect more comprehensive user preference. In addition, Cheng et al. [10] proposed a multi-modal aspect-aware topic model (MATM) on text reviews and item images to model users' preferences and items' features from different aspects. Although these efforts achieved significant progress, the multi-modal information lying in the review or description is usually informal, unstructured, and noisy, which largely limits the model performance. To address this issue, Hou et al. [19] proposed to extract the attribute information from the item image, and utilize the attention mechanism to find the explanations for recommendation results. Nevertheless, due to the limited expressive capability of the item images in reflecting the item attributes, this method still suffers from the sub-optimal performance. Beyond that, in this work, we exploit the explicit item attributes to boost the recommendation performance.

5.2.2   Graph-Based Recommendation

Early graph-based recommendation aims to construct a simple user-item interaction graph and exploit it to infer user preferences, such as ItemRank [12] and BiRank [36]. Specifically, ItemRank adopts a random walk-based algorithm to predict item scores for the target users and BiRank iteratively assigns scores to vertices and finally converges to a unique stationary ranking. Later, some researchers [4, 40] attempted to use the connection of entities in the knowledge graph for embedding learning to improve the recommendation performance. For example, Zhang et al. [40] adopted TransR to learn item embeddings under the guidance of structural information in the knowledge graph. One key limitation of these studies is that they overlook the higher-order connectivity [34] in the knowledge graph, which is helpful for mining the potential interests of users. Therefore, some researchers proposed to perform iterative propagation over the entire knowledge graph for user preference modeling.

For instance, Wang et al. [33, 35] applied graph neural networks to achieve embedding propagation on the knowledge graph. In addition, He et al. [17] developed LightGCN, which improves model efficiency without sacrificing accuracy by removing the nonlinear activation and feature transformation from NGCF [35].

Recent research [1, 6, 20, 25] propose to incorporate the semantic information into the graph learning. Most of them propose to construct a unified graph of users, items, and attributes and model their relationships through the information propagation. For example, Wang et al. [34] proposed KGAT, which jointly considers user-item and item-attribute interactions in the information propagation process. Ai et al. [1] introduced the knowledge graph to organize items' attributes, and designed a path-based soft matching algorithm to generate corresponding explanations. Similarly, Chen et al. [6] modeled the connectivity between different entities (i.e., user, item and attribute) in the knowledge graph via GCN.

Although these graph-based recommendation methods can learn the relationships between different entities, they overlook different relation types between entities. In fact, different relation types indicate different correlations among entities and contribute differently when conveying the user preference. Therefore, in this work, we propose to learn the user preference with a newly designed relation-aware attentive propagation method, which learns the different contributions of different nodes in the graph during the information propagation based on relation types.

5.3 Problem Formulation

Suppose we have a set of users $\mathcal{U} = \{u_i\}_{i=1}^{N_u}$, and a set of fashion items $\mathcal{P} = \{p_j\}_{j=1}^{N_p}$, where N_u and N_p are the total numbers of users and items, respectively. Meanwhile, we have a set of possible attribute values (e.g., *red color* and *wool material*) possessed by items, denoted as $\mathcal{A} = \{a_m\}_{m=1}^{N_a}$, where N_a is the total number of attribute values. Notably, each fashion item is associated with an image and a few attribute values of $\mathcal{A}$. Let $\mathcal{V} = \{v_j\}_{j=1}^{N_p}$ denote the set of item images, where v_j is the image of item p_j.

On the one hand, based on the user-item historical interactions (e.g., user u_i purchases item p_j) and the item-attribute association relations (e.g., item p_j has the attribute a_l), we can derive a user-item-attribute tripartite graph $\mathcal{G}_a = (\mathcal{E}_a, \mathcal{R}_a)$. In particular, the node set $\mathcal{E}_a = \mathcal{U} \bigcup \mathcal{P} \bigcup \mathcal{A}$ denotes the set of entities, including user entities, item entities and attribute entities, while $\mathcal{R}_a = \{r_1, \ldots, r_{2(F+1)}\}$ refers to the set of relation types held by these entities, which includes one type of user-item interaction relation as well as F types of item-attribute association relations, and their reverse ones. For instance, the relation *"InteractBy"* is the reverse form of the relation *"Interact"*, while *"IsColorOf"* is the reverse one of the relation *"HasColorOf"*. The motivation that we incorporate the reverse relations is to allow the information propagation in both directions. In this manner, each item entity can absorb knowledge from its attribute entities, while the attribute entity can also distill information from its related item entities, which benefits the latent representation learning of entities. For ease of the following method description, similar to previous works, we formulate the user-item-attribute graph $\mathcal{G}_a$ as a set of triplets $\mathcal{T} = \{(h, r, t)|h, t \in \mathcal{E}_a, r \in \mathcal{R}_a\}$, where each triplet (h, r, t) indicates that there is a relation of the type r from the head entity h to the tail entity t. Here we give some intuitive triplet examples, such as $(user1, bought, item1)$, $(item1, HasColorOf, black)$, and $(black, IsColorOf, item1)$.

On the other hand, based upon the user-item historical interactions, besides the relation-oriented graph $\mathcal{G}_a$, we can also derive a vision-oriented graph $\mathcal{G}_v = (\mathcal{E}_v, \mathcal{R}_v)$ to characterize the user-item interactions from the vision aspect. In particular, $\mathcal{E}_v = \mathcal{U} \bigcup \mathcal{V}$ denotes the set of entities, including the user entities and image entities, while $\mathcal{R}_v$ contains only one relation type, i.e., the interaction relation between the user and the item image. The reason why we do not incorporate the reversed relation in the vision-oriented graph is the concern that the user's visual preferences should be derived by summarizing the objective visual features of items that the user interacted before, while the bidirectional relation that allows the bidirectional information propagation may hurt such objectivity. Similar to $\mathcal{G}_a$, the vision-oriented graph $\mathcal{G}_v$ can be represented as a triplet set $\mathcal{Z} = \{(u, r', v)|u \in \mathcal{U}, v \in \mathcal{V}\}$, where each triplet (u, r', v) represents that the user u once interacted with an item that has the image of v, and $r' = interact$.

Goal. In this work, based on these two graphs, i.e., $\mathcal{G}_a$ and $\mathcal{G}_v$, we aim to train a model $\mathcal{F}$ that is able to predict the preference score of an arbitrary user u to an arbitrary item p as follows,

$$\hat{y}_{u,p} = \mathcal{F}(u, p|\mathcal{G}_a, \mathcal{G}_v), \tag{5.1}$$

where $\hat{y}_{u,p}$ denotes the predicted preference score of the user u towards the item p.

5.4   MM-FRec

In this section, we present the proposed MM-FRec, as shown in Fig. 5.1, which consists of three key components: (1) *Attribute-enhanced Latent Representation Learning*, (2) *Visual Representation Learning*, and (3) *Multi-modal Enhanced Preference Modeling*. The first component works on learning the attribute-enhanced latent representations for users and items over the relation-oriented tripartite graph. The second component targets on investigating the visual representations of users and items based upon the vision-oriented bipartite graph. The last component aims to learn the multi-modal enhanced user preference based on the attribute-enhanced representations and visual representations of users and items.

5.4.1   Attribute-Enhanced Latent Representation Learning

This component consists of two key modules: *Tripartite Graph Embedding* and *Relation-aware Attentive Propagation*. The former is designed to initialize the entity and relation embeddings of the user-item-attribute tripartite graph at each run. The latter is devised to adaptively perform the information propagation over the tripartite graph to derive the attribute-enhanced latent representation of users and items, where we assign the weights to neighbor nodes according to not only the head/tail nodes of a triplet but also their corresponding relation type. In particular, to deal with the unique attributes, we borrow the idea of multi-task learning and fulfil the relation-aware attention mechanism through a shared network and multiple relation-specific networks.

Tripartite Graph Embedding. Knowledge graph embedding has been proven to be an effective way to parameterize entities and relations as vector representations by preserving the graph structure, which has been used in various knowledge enhanced recommendation methods [1, 6]. Therefore, in our context, to learn the entity and relation embeddings of the user-item-attribute tripartite graph, we resort to the widely-used knowledge graph embedding method TransR [23].

Specifically, for each triplet (h, r, t) in $\mathcal{T}$, i.e., the relation-oriented graph, we first project the head entity h and tail entity t into the corresponding relation space as follows,

$$\mathbf{e}_{h_r} = \mathbf{M}_r \mathbf{e}_h, \quad \mathbf{e}_{t_r} = \mathbf{M}_r \mathbf{e}_t, \tag{5.2}$$

where $\mathbf{e}_h \in \mathbb{R}^d$ and $\mathbf{e}_t \in \mathbb{R}^d$ represent the to-be-learned original embeddings of entities h and t, respectively. d is the embedding dimension. $\mathbf{M}_r \in \mathbb{R}^{d \times d}$ denotes the corresponding projection matrix of the relation type $r \in \mathcal{R}_a$. $\mathbf{e}_{h_r} \in \mathbb{R}^d$ and $\mathbf{e}_{t_r} \in \mathbb{R}^d$ are the projected representations of the head and tail entities in the latent space of the relation type r, respectively.

We then build a training set $Q = \{(h, r, t, t') | (h, r, t) \in \mathcal{T}, (h, r, t') \notin \mathcal{T}\}$, where $t' \neq t$ is a negative tail entity regarding the head entity h and relation r, randomly sampled from the whole entity set $\mathcal{E}_a$. Based on the training set, we use the bayesian ranking loss function for entity embedding of the relation-oriented graph as follows,

$$\begin{cases} \mathcal{L}_{GE} = \sum_{(h,r,t,t') \in Q} \ln \sigma \left(g(h, r, t) - g\left(h, r, t'\right) \right) + \lambda \|\Theta_1\|_F^2, \\ g(h, r, t) = \left\| \mathbf{e}_{h_r} + \mathbf{e}_r - \mathbf{e}_{t_r} \right\|_2^2, \end{cases} \tag{5.3}$$

where $\sigma(\cdot)$ is the sigmoid function, and $\mathbf{e}_r \in \mathbb{R}^d$ represents the to-be-learned embedding of relation type r. λ is a non-negative hyperparameter and Θ_1 refers to the parameters $\{\mathbf{M}_{r_1}, \mathbf{M}_{r_2}, \ldots, \mathbf{M}_{r_{2(F+1)}}\}$ of the graph embedding. The L_2 norm regularization on Θ_1 is introduced to prevent overfitting. Essentially, the lower the value of $g(h, r, t)$, the more likely the triplet exists in the graph. Then the BPR loss function encourages the positive triplet (h, r, t) to get a lower value than the negative ones. Ultimately, by optimizing the above objective function, we can obtain the embeddings for all entities (i.e., user entities, item entities, and attribute entities) and relations.

Relation-aware Attentive Propagation. After obtaining the entity embedding and relation embedding of the relation-oriented graph $\mathcal{G}_a$, we proceed to the information propagation over it to uncover the user's and item's attribute-enhanced latent representation. Intuitively, one user tends to prefer items that share similar properties or have been bought

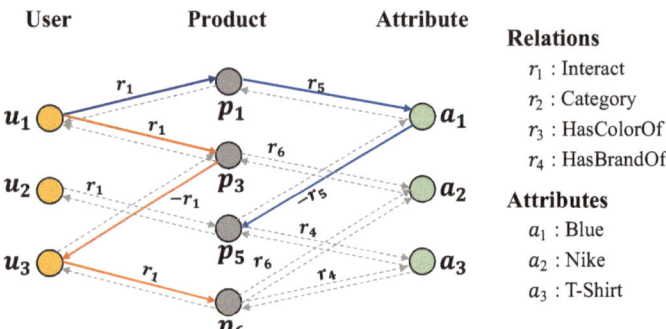

Fig. 5.2 Illustration of high-order connectivities

by users with similar tastes, which can be reflected by the high-order connectivities in the relation-oriented graph. As illustrated by Fig. 5.2, based on the high-order connectivities: $u_1 \xrightarrow{r_1} p_1 \xrightarrow{r_5} a_1 (black) \xrightarrow{-r_5} p_5$ and $u_1 \xrightarrow{r_1} p_3 \xrightarrow{-r_1} u_3 \xrightarrow{r_1} p_6$, where $r_1, r_5, -r_1$ and $-r_5$ refer to the relations *"Interact"*, *"HasColorOf"*, *"InteractBy"* and *"IsColorOf"*, respectively, we can infer that the user u_1 may be interested in items p_5 and p_6. In light of this, we argue that it is promising to explore the high-order connectivity among entities in the relation-oriented graph to uncover the user's attribute-enhanced latent preference. Toward this end, similar to existing works, we explore the high-order connectivities among entities by conducting multiple iterations of the 1-order (i.e., one-hop) information propagation. Therefore, we describe the 1-order information propagation as an example, while the high-order information propagation can be derived iteratively.

During the 1-order information propagation, each node refines its embedding by comprehensively summarizing information from its ego-network [28], i.e., its one-hop neighbors. Formally, we regard each node as the head entity h, and define its ego-network as $\mathcal{T}_h = \{(h, r, t) | (h, r, t) \in \mathcal{T}\}$. Moreover, as a matter of fact, different attributes contribute differently in expressing the item's features, and different items also act differently in characterizing the user's latent preference. Consequently, we adopt the attention mechanism to dynamically assign the confidence of entity nodes in one's ego-network during the information propagation. Notably, even for the same head-tail entity pair, e.g., $(item1, Middle\_length)$, there can be various relations, like "sleeve_length" and "dress_length", and the specific relation should also affect the confidence of the neighbor node (i.e., tail entity) influencing the central head entity. Accordingly, we propose to utilize all the three elements of each triplet to measure the confidence of each tail entity in the ego-network towards the central head entity representation learning.

Different from existing studies that measure the triplet confidence with the simple linear transformation, we employ the neural networks to capture the latent complex interactions among the three elements of a triplet. Moreover, considering the unique attributes that being possessed by limited samples, we adopt the multi-task learning framework to learn the

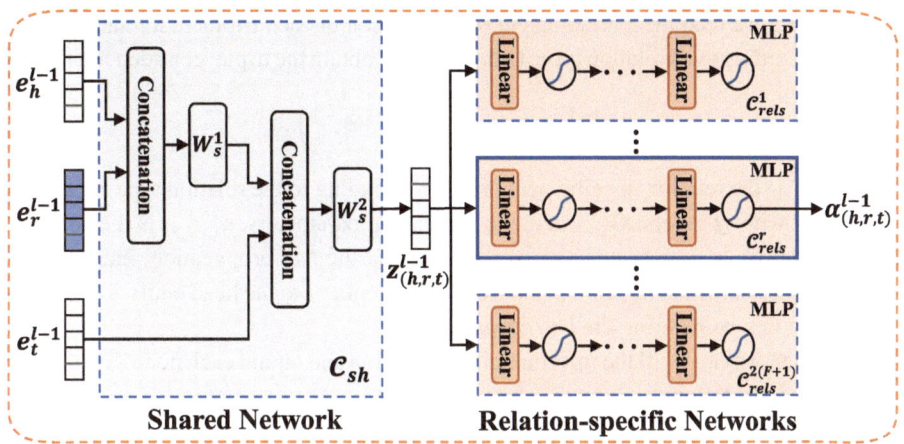

Fig. 5.3 Deep multi-task learning based confidence learning for each triplet

triplet confidence. Specifically, we regard the confidence assignment of triplets with different relation types as a set of correlated tasks. The underlying philosophy is that there should be some common knowledge on the triplet confidence assignment regardless of relation types. In particular, as illustrated in Fig. 5.3, we employ a shared network and multiple relation-specific networks to learn the confidence for each triplet in the node's ego-network. The shared network works on learning the latent representation of the triplet that encodes the common knowledge on the triplet confidence assignment, while the multiple relation-specific networks target at learning the relation-specific triplet confidence. Both shared network and relation-specific networks are constructed via the multi-layer perceptron (MLP).

Specifically, we devise the shared network with two fully-connected layers, where the first layer aims to project the head and the relation embeddings into the common tail entity space, while the second one serves to learn the latent interaction among the three elements in the common space. Formally, we have,

$$
\begin{aligned}
\mathbf{z}_{(h,r,t)}^{l-1} &= C_{sh}(\mathbf{e}_h^{l-1}, \mathbf{e}_r^{l-1}, \mathbf{e}_t^{l-1}) \\
&= \tau \left(\mathbf{W}_s^2 \left(\tau \left(\mathbf{W}_s^1 \left(\mathbf{e}_h^{l-1} \| \mathbf{e}_r^{l-1} \right) + \mathbf{b}_s^1 \right) \| \mathbf{e}_t^{l-1} \right) + \mathbf{b}_s^2 \right),
\end{aligned}
\tag{5.4}
$$

where $\|$ is the concatenation operation. $\mathbf{e}_h^{l-1} \in \mathbb{R}^d$, $\mathbf{e}_r^{l-1} \in \mathbb{R}^d$, and $\mathbf{e}_t^{l-1} \in \mathbb{R}^d$ denote the embedding of the head entity, relation, tail entity of the triplet (h, r, t) during the l-th propagation, $l = \{1, 2, \ldots, L\}$, respectively. L is the total number of the propagation iterations. In particular, $\mathbf{e}_h^0$, $\mathbf{e}_r^0$, and $\mathbf{e}_t^0$ are obtained by the aforementioned relation-oriented graph embedding. $\mathbf{W}_s^1 \in \mathbb{R}^{2d \times d}$, and $\mathbf{W}_s^2 \in \mathbb{R}^{2d \times d}$ are the weight matrices of the shared neural network C_{sh}, while $\mathbf{b}_s^1 \in \mathbb{R}^d$, and $\mathbf{b}_s^2 \in \mathbb{R}^d$ are the bias vectors of C_{sh}. $\tau(\cdot)$ refers to the tanh active function [32]. $\mathbf{z}_{(h,r,t)}^{l-1}$ denotes the learned latent representation of the triplet by the shared network in the l-th propagation.

Afterward, we feed the learned latent representation of each triplet to a relation-specific network according to its relation type. Ultimately, we obtain the triplet confidence as follows,

$$\alpha_{(h,r,t)}^{l-1} = C_{rels}^r \left(\mathbf{z}_{(h,r,t)}^{l-1} | \Theta_r^{l-1} \right), \tag{5.5}$$

where C_{rels}^r is the relation-specific network corresponding to the relation type r, composed by an MLP with Q layers. Θ_r^{l-1} is the to-be-learned parameters. $\alpha_{(h,r,t)}^{l-1}$ is the confidence for the triplet (h, r, t) in the ego-network $\mathcal{T}_h$ during the l-th propagation, indicating how much information would be propagated from the tail entity t to the head entity h conditioned on the relation type r during the l-th propagation.

Thereafter, we can fulfil the information propagation and obtain each node's ego-network representation as follows,

$$\mathbf{e}_{\mathcal{T}_h}^{l-1} = \sum_{(h,r,t) \in \mathcal{T}_h} \alpha_{(h,r,t)}^{l-1} \cdot \mathbf{e}_t^{l-1}. \tag{5.6}$$

Ultimately, due to the remarkable capability of the Bi-Interaction aggregator function in aggregating information [34], we adopt it to aggregate the representations of the node self and its ego-network as the refined node representation after the l-th propagation as follows,

$$\mathbf{e}_h^l = \omega \left(\mathbf{W}_1^l (\mathbf{e}_h^{l-1} + \mathbf{e}_{\mathcal{T}_h}^{l-1}) \right) + \omega \left(\mathbf{W}_2^l (\mathbf{e}_h^{l-1} \odot \mathbf{e}_{\mathcal{T}_h}^{l-1}) \right), \tag{5.7}$$

where $\omega(\cdot)$ is the LeakyReLU active function, and $\odot$ denotes the element-wise multiplication operation. $\mathbf{W}_1^l$ and $\mathbf{W}_2^l \in \mathbb{R}^{d_l \times d_{l-1}}$ are the weight matrices, where $d_l = d_{l-1}/2$, and $d_0 = d$.

After L propagation iterations, we can derive L embeddings for each entity h correspondingly, i.e., $\{\mathbf{e}_h^1, \mathbf{e}_h^2, \dots, \mathbf{e}_h^L\}$. Ultimately, we concatenate all these representations with the initial embedding into a single vector $\mathbf{e}_h^a \in \mathbb{R}^{(d_0 + \cdots + d_L)}$ to get the final representations of the entity h as follows,

$$\mathbf{e}_h^a = \mathbf{e}_h^0 \| \mathbf{e}_h^1 \| \cdots \| \mathbf{e}_h^L. \tag{5.8}$$

As the head entity h can be a user entity, an item entity, or an attribute entity, for clarity, we further use $\mathbf{e}_u^a$, $\mathbf{e}_p^a$, and $\mathbf{e}_a^a$ to denote the final latent representation of the user entity u, the item entity p, the attribute entity a, derived according to Eq. (5.8), respectively.

5.4.2 Visual Representation Learning

Apart from the attribute-enhanced latent representation learning, we also introduce the visual representation learning to model the user's visual preference. In particular, we first employ the pre-trained visual representation learning model ResNet50 [15], which has been widely used for visual feature extraction [13, 24, 27, 31], to generate the initial visual representation of each item image. Let $\tilde{\mathbf{e}}_p \in \mathbb{R}^{d_i}$ be the initial visual representation of the item p, where $d_i = 2048$. We then introduce an MLP with two fully-connected layers to reduce the dimension of the representation of each item image and realize the fine-tuning as follows,

$$
\begin{aligned}
\mathbf{e}_p^v &= \mathcal{H}^v \left(\tilde{\mathbf{e}}_p \mid \mathbf{\Theta}^v \right) \\
&= \tau \left(\mathbf{W}_v^2 \left(\tau \left(\mathbf{W}_v^1 \tilde{\mathbf{e}}_p + \mathbf{b}_v^1 \right) \right) + \mathbf{b}_v^2 \right)
\end{aligned}
\tag{5.9}
$$

where $\mathbf{e}_p^v \in \mathbb{R}^{d_v}$ is the fine-tuned visual representation of the item p, and $\mathbf{\Theta}^v$ refers to the parameters of the MLP $\mathcal{H}^v$. Specifically, $\mathbf{W}_v^1 \in \mathbb{R}^{\frac{d_j}{4} \times d_i}$ and $\mathbf{W}_v^2 \in \mathbb{R}^{d_v \times \frac{d_j}{4}}$ are the weight matrices of the MLP $\mathcal{H}^v$, while $\mathbf{b}_v^1 \in \mathbb{R}^{\frac{d_j}{4}}$ and $\mathbf{b}_v^2 \in \mathbb{R}^{d_v}$ are the bias vectors of $\mathcal{H}^v$. $\tau(\cdot)$ refers to the tanh active function. To facilitate the following multi-modal enhanced preference learning, we make the dimension of visual representation equal to that of the attribute-enhanced latent representation $\mathbf{e}_p^a$, i.e., $d_v = (L+1)d$.

Due to the lack of direct visual cues of the user, we resort to the visual representations of images in the user entity's ego-network in graph $\mathcal{G}_v$ to obtain the user's visual embedding. In particular, we first define the user's visual ego-network as the set of triplets whose head entities are the user entity u in $\mathcal{Z}$, denoted as $\mathcal{Z}_u = \{(u, r', v) \mid (u, r', v) \in \mathcal{Z}\}$. Thereafter, we obtain the user's visual representation as follows,

$$
\mathbf{e}_u^v = \frac{1}{|\mathcal{Z}_u|} \sum_{(u,r,v) \in \mathcal{Z}_u} \mathbf{e}_p^v,
\tag{5.10}
$$

where v is the image of the item p, and $\mathbf{e}_u^v \in \mathbb{R}^{d_v}$ is the visual embedding for user u.

5.4.3 Multi-modal Enhanced Preference Learning

To comprehensively capture the user's preference, we fuse the user's and item's visual representation with their corresponding attribute-enhanced latent representations. Specifically, we have,

$$
\mathbf{e}_u^f = \mathbf{e}_u^a \odot \mathbf{e}_u^v, \quad \mathbf{e}_p^f = \mathbf{e}_p^a \odot \mathbf{e}_p^v,
\tag{5.11}
$$

where $\odot$ denotes the element-wise multiplication operation. $\mathbf{e}_u^f$ and $\mathbf{e}_p^f$ denote the multi-modal enhanced representation of the user u and item p, respectively.

Ultimately, we employ the inner product between the multi-modal enhanced representations of a user u and an item p to measure the user u's preference to item p as follows,

$$
\hat{y}_{u,p} = (\mathbf{e}_u^f)^T \mathbf{e}_p^f,
\tag{5.12}
$$

where $\hat{y}_{u,p}$ represents the predicted preference of the user u to item p.

Intuitively, the visual image and semantic attributes characterize the same fashion item, and should share certain latent consistency. In other words, the visual representation and the attribute-enhanced latent representation of the same item should be more similar than that of different items. Therefore, we utilize the contrastive loss as a visual-semantic consistency regularizer, which has shown superior performance in characterizing the inter-modal consistency and bridges the semantic gap between the low-level visual clues and high-level attribute semantics [38]. Formally, we have,

$$\mathcal{L}_{VSE} = \sum_{i \neq j} \left(\max \left\{ 0, \beta - \cos \left(e^a_{p_i}, e^v_{p_i} \right) + \cos \left(e^a_{p_i}, e^v_{p_j} \right) \right\} \right.$$
$$\left. + \max \left\{ 0, \beta - \cos \left(e^v_{p_i}, e^a_{p_i} \right) + \cos \left(e^v_{p_i}, e^a_{p_j} \right) \right\} \right),$$

(5.13)

where β is a margin hyperparameter and $\cos(\cdot, \cdot)$ refers to the cosine similarity operator. $e^a_{p_i}$ and $e^v_{p_i}$ are the attribute-enhanced latent representation and the visual embedding of the item entity p_i, respectively.

Algorithm 5.1 The training procedure of MM-FRec

Input: relation-oriented graph $\mathcal{G}_a$, vision-oriented graph $\mathcal{G}_v$, training set $\mathcal{Q}$ for the tripartite graph embedding and training set $\mathcal{D}$ for the user preference learning, image set $\mathcal{V}$, and mini-batch size m. Hyperparameters: $\{\lambda, \beta, \mu, \eta\}$
Output: Parameters $\{\Theta_1, \Theta_2\}$.

1: **Initialization**
2: Initialize all parameters: $\{\Theta_1, \Theta_2\}$.
3: **repeat**
4:     **while** an epoch is not end **do**
5:         Randomly sample a mini-batch of (h, r, t, t')'s from the training set $\mathcal{Q}$.
6:         Update Θ_1 by optimizing the loss function in Eq. (5.3).
7:     **end while**
8:     **while** an epoch is not end **do**
9:         Randomly sample a mini-batch of (u, p, p')'s from the training set $\mathcal{D}$.
10:        Update Θ_2 by optimizing the loss function in Eq. (5.16).
11:     **end while**
12: **until** MM-FRec converges

Optimization. For optimization, we adopt the widely-used BPR loss, which has been proven to be powerful in the pairwise implicit preferences modeling [3, 7, 14]. Specifically, we first construct the following training set:

$$\mathcal{D} = \left\{ (u, p, p') \mid u \in \mathcal{U}, \ p, p' \in \mathcal{P}, (u, r, p) \in \mathcal{T}, (u, r, p') \notin \mathcal{T} \right\},$$

(5.14)

where the negative item p' was randomly sampled from $\mathcal{T}$, and r refers to the "*interact*" relation. We then have the BPR loss function as follows,

$$\mathcal{L}_{BPR} = - \sum_{(u, p, p') \in \mathcal{D}} \ln \sigma \left(\widehat{y}_{(u, p)} - \widehat{y}_{(u, p')} \right) + \mu \|\Theta_2\|^2_F,$$

(5.15)

where μ is the non-negative hyperparameter. The last term is introduced to avoid overfitting and Θ_2 refers to the set of model parameters (i.e., $\{\mathbf{W}_s^1, \mathbf{b}_s^1, \mathbf{W}_s^2, \mathbf{b}_s^2, \mathbf{W}_1^l, \mathbf{W}_2^l, \Theta_r^l, \Theta^v\}$). Intuitively, we expect that a user's preference to the positive item should be larger than that to the negative item.

Finally, we have the following objective function,

$$\mathcal{L} = \mathcal{L}_{BPR} + \eta \mathcal{L}_{VSE}, \tag{5.16}$$

where η is the non-negative hyperparameter.

During the training phase, similar to KGAT [34], we optimize $\mathcal{L}$ and $\mathcal{L}_{GE}$ alternatively. The overall procedure of the optimization is briefly summarized in Algorithm 5.1. As can be seen, in each run, we first optimize the entity embeddings of the user-item-attribute graph and the projection matrices $\mathbf{M}_r$'s of TransR according to Eq. (5.3), and then optimize the other parameters of our model, including those in the relation-aware attentive propagation, visual representation learning, and multi-modal enhanced preference learning, according to Eq. (5.16).

5.5 Experiments

In this section, we first present the dataset and experimental setting, and then introduce the extensive experiments that we conducted on a real-world dataset by answering the following research questions:

1. Does the proposed MM-FRec outperform state-of-the-art methods?
2. What is the effect of each component on the performance of our MM-FRec?
3. How is the sensitivity of our model?
4. What is the intuitive performance of our model?

5.5.1 Dataset

Towards the fashion item recommendation, there have been several public benchmark datasets, such as POG [8], Amazon [19], and Tradesy [16]. Nevertheless, most of them (e.g., Tradesy) lack the attribute information of items. Although POG provides a set of attributes for each item, it lacks the corresponding attribute types. In addition, the Amazon dataset only contains the item's size attribute and color attribute. These deficiencies lead to the limitations of their applications in our context. Therefore, we resorted to the dataset IQON3000 [30], which was originally constructed for the personalized outfit recommendation. To be specific, IQON3000 comprises 308,747 outfits of 3,568 users, involving 672,335 items. Each item is associated with a visual image, and several fine-grained attributes. In par-

Table 5.1 Statistics of IQON10

#Users	#Items	#User-item
3,568	23,363	459,146
#Attribute types	#Attribute values	#Item-attribute
11	3,019	153,020

Table 5.2 Attribute distribution of different types in IQON10

Attribute type	#Triplets	Attribute type	#Triplets
Color	46,292	Pattern	2,371
Price	23,363	Design	1,624
Brand	23,360	Heel	929
Category	23,334	Dress_length	550
Variety	22,929	Sleeve_length	518
Material	7,750		

ticular, in total, there are 11 attribute types, including *color, price, brand, category, variety, material, pattern, design, heel, dress_length*, and *sleeve_length*.

To adapt IQON3000 to our task of fashion item recommendation, we treated items in an outfit composed by a user as the positive items for this user. We argue that if a user publicly shares an outfit, then he/she should prefer the composing items of the outfit. To ensure the quality of the dataset, we filtered out the items that are preferred by less than 10 users. Ultimately, the final derived dataset, named as IQON10, comprises 3,568 users, 23,363 fashion items and 459,146 user-item interactions. In addition, there are 3,019 attribute values, and 153,020 item-attribute association relations. Table 5.1 summarizes the statistics of our dataset. In addition, Table 5.2 shows the detailed statistics for different attribute types. As can be seen, the number of the triplets with different attribute types varies in a large range. Concretely, a large number of items have the "color" attribute, while limited items have the attributes of "Dress_length" and "Sleeve_length", which confirms the existence of the unique attributes in practice.

5.5.2   Experimental Settings

In this part, we present the evaluation metrics, baselines, and implementation details.

Evaluation Metrics. We evaluated our method with the task of Top-K recommendation [18, 41], and adopted three widely-used evaluation metrics [5, 21]: $Recall@K$, $NDCG@K$ and $Precision@K$. In this work, we set $K = 20, 50, 100$, respectively.

Baselines. To prove the effectiveness of MM-FRec, we compared the proposed MM-FRec with five state-of-the-art baselines.

- **CFKG** [1]: This model only relies on the user-item-attribute graph to model the user's preference, which adopts TransE [2] to embed the heterogeneous entities in the user-item-attribute graph.
- **KGAT** [34]: This model explores the high-order connectivity in the collaborative knowledge graph, i.e., the user-item-attribute graph in our context, with graph neural network and the attention mechanism. Notably, this method overlooks the user's visual preference modeling.
- **JNSKR** [6]: This method uses an efficient Non-Sampling (NS) optimization algorithm for the user-item-attribute graph embedding, which is able to train the model by the whole training data with a low time complexity. This method also ignores the visual modality of items.
- **VBPR** [16]: This is a multi-modal based model, which incorporates the visual modality of fashion items to enhance the factorization-based user preference modeling.
- **SAERS** [19]: This is also a multi-modal based approach that characterizes the preference of users from both the visual and semantic attribute perspectives. Notably, since our IQON10 contains specific attributes of fashion items, we directly adopted these ground-truth attribute labels without pre-training an additional attribute extraction network in the reproduction of this model.

Implementation Details. We randomly sampled 80%, 10%, and 10% of the positive user-item interactions in our dataset to form the training set, validation set and testing set, respectively. Regarding the user-item-attribute graph, there are 24 relation types, corresponding to the user-item interaction relation type and 11 attribute types, as well as their reverse ones. Accordingly, for the relation-aware attentive propagation, apart from the shared network, we deployed 24 independent relation-specific networks for the confidence assignment toward nodes in the ego-network. The size of item images is unified to 150×150. Pertaining to the optimization, we utilized the adaptive moment estimation method (Adam) [22], and all the parameters in the neural networks are initialized by Xavier initializer [11]. The grid search strategy is adopted to determine the optimal values for the regularization hyperparameters λ and μ among values $\{10^r | r \in \{-5, \cdots, -2\}\}$. In addition, the learning rate lr, the embedding dimension d, the number of layers in the relation-specific network Q, the number of iterations of information propagation L, the margin of contrastive loss β, and the hyperparameter for contrastive loss η are searched in $[0.001, 0.0005, 0.0001]$, $[16, 32, 64]$, $[1, 2, 3, 4]$, $[1, 2, 3, 4]$, $\{10^r | r \in \{-5, \cdots, 1\}\}$, and $[0.1, 0.5, 1, 5, 10]$, respectively. Finally, we empirically found that the proposed model achieves the optimal performance with regularization hyperparameters $\lambda = \mu = 10^{-5}$, the learning rate $lr = 0.0005$, the embedding dimension $d = 64$, the number of layers in the relation-specific network $Q = 3$, the number of iterations of information propagation $L = 3$, the margin of contrastive loss $\beta = 10^{-3}$, and the hyperparameter for contrastive loss $\eta = 5$.

5.5.3   On Model Comparison

Table 5.3 shows the performance of different methods in the Top-K recommendation task. To make a fair comparison with baseline methods that only adopt the attribute information of each item, we also derived A-FRec from our MM-FRec by removing the visual modality. From Table 5.3, we have the following observations. (1) Our MM-FRec outperforms all the other baselines in the two recommendation tasks on almost all evaluation metrics. In particular, compared with the best baseline, MM-FRec achieves a significant improvement of 28.40%, 20.22%, 24.82% in Precision@20, Precision@50, and NDCG@100, respectively. This demonstrates the superiority of our MM-FRec and reflects that our model tends to rank the positive items at the top places, as compared to the baseline methods. (2) Notably, we also conducted a pairwise significant test, and the results verify that all the improvements of MM-FRec are statistically significant with p-value < 0.01. (3) MM-FRec surpasses A-FRec in all cases, confirming the advantage of considering the visual content of items in the fashion item recommendation task. And (4) even removing the visual modality, our A-FRec still outperforms KGAT and JNSKR regarding almost all metrics, which indicates the effectiveness of our relation-aware attentive propagation.

5.5.4   On Ablation Study

In the ablation study, we investigated the three key components of our model: relation-aware attention mechanism, visual-semantic consistency regularizer, and multi-modal fusion manner. **Effect of Relation-aware Attention Mechanism.** As aforementioned, the proposed relation-aware attention mechanism simultaneously considers the three elements of each triple, i.e., the head entity, the relation, and the tail entity, to assign the node confidence. Meanwhile, considering the existence of unique attributes, we devised a shared network and

Table 5.3 Performance comparison among different approaches. The numbers in bold indicate the best results, while those underlined refer to the second best results. * denotes that the p-value of the significant test between our result and the best baseline result is less than 0.01. "R": "Recall"; "P": "Precision"; "Improv.": "Performance improvement"

Model	R@20	NDCG@20	P@20	R@50	NDCG@50	P@50	R@100	NDCG@100	P@100
CFKG	0.1860	0.0940	0.0172	0.2837	0.1154	0.0122	<u>0.3589</u>	0.1383	0.0097
KGAT	0.1643	0.0900	0.0199	0.2287	0.1011	0.0134	0.3081	0.1204	0.0109
JNSKR	<u>0.2073</u>	0.1005	0.0116	<u>0.2850</u>	<u>0.1462</u>	0.0083	0.3126	<u>0.1543</u>	0.0073
VBPR	0.0932	0.0505	0.0092	0.1243	0.0769	0.0062	0.1506	0.0821	0.0045
SAERS	0.1863	<u>0.1252</u>	<u>0.0243</u>	0.2202	0.1373	<u>0.0178</u>	0.2479	0.1460	**0.0167**
A-FRec	0.2071	0.1201	0.0274	0.2941	0.1441	0.0192	0.3662	0.1634	0.0146
MM-FRec	**0.2317***	**0.1486***	**0.0312***	**0.3157***	**0.1732***	**0.0214***	**0.3857***	**0.1926***	<u>0.0161*</u>
Improv.	11.77%↑	18.69%↑	28.40%↑	10.77%↑	18.47%↑	20.22%↑	7.47%↑	24.82%↑	–

multiple relation-specific networks for the attention calculation. Accordingly, to comprehensively evaluate the relation-aware attention mechanism in our model, we introduced the
following five variant models of our MM-FRec:

- **MM-FRec-NoRA**: The whole relation-aware attention mechanism is disabled.
- **MM-FRec-NoSH**: The shared network of our model is removed.
- **MM-FRec-NoRS**: The relation-specific networks of our model are deleted.
- **MM-FRec-Linear**: To justify the benefit of using neural networks, we introduced this
  method by replacing the neural networks with the linear transformations.
- **MM-FRec-NoR**: To verify the necessity of incorporating the relation element of a triplet
  in the confidence assignment, we removed the $\mathbf{e}_r^{l-1}$ from Eq. (5.4).

Table 5.4 shows the performance of our MM-FRec and A-FRec with different attention configurations. From this table, we have the following observations. (1) MM-FRec-
NoRA and A-FRec-NoRA perform worse than our MM-FRec and A-FRec, respectively.
This verifies the necessity of the dynamic triplet confidence assignment in the information
propagation over the user-item-attribute tripartite graph. (2) MM-FRec and A-FRec outperform MM-FRec-NoSH (MM-FRec-NoRS) and A-FRec-NoSH (A-FRec-NoRS), respectively. This confirms the benefit of incorporating the deep multi-task learning strategy to
handle the triplet confidence assignment of various attributes. (3) MM-FRec and A-FRec
surpass MM-FRec-NoR and A-FRec-NoR, respectively. This verifies the necessity of introducing relation element in the confidence assignment during information propagation. (4)
MM-FRec and A-FRec outperform MM-FRec-Linear and A-FRec-Linear, respectively. This

Table 5.4 Effect of the relation-aware attention mechanism on our model

Model	Recall@20	NDCG@20	Precision@20
A-FRec-NoRA	0.1850	0.1068	0.0247
A-FRec-NoSH	0.1988	0.1169	0.0264
A-FRec-NoRS	0.1973	0.1148	0.0263
A-FRec-Linear	0.1754	0.0983	0.0206
A-FRec-NoR	0.1981	0.1168	0.0263
A-FRec	**0.2056**	**0.1205**	**0.0268**
MM-FRec-NoRA	0.2128	0.1327	0.0282
MM-FRec-NoSH	0.2281	0.1426	0.0302
MM-FRec-NoRS	0.2310	0.1468	0.0308
MM-FRec-Linear	0.1998	0.1210	0.0272
MM-FRec-NoR	0.2085	0.1238	0.0272
MM-FRec	**0.2317**	**0.1486**	**0.0312**

Table 5.5 Effect of visual-semantic consistency regularizer on our model

Model	Recall@20	NDCG@20	Precision@20
MM-FRec-NoVSE	0.2215	0.1312	0.0299
MM-FRec	**0.2317**	**0.1486**	**0.0312**

Table 5.6 Performance of different modality fusion methods

Model	Recall@20	NDCG@20	Precision@20
MM-FRec$_+$	0.2018	0.1190	0.0279
MM-FRec$_{||}$	0.2106	0.1289	0.0273
MM-FRec	**0.2317**	**0.1486**	**0.0312**

proves the effectiveness of neural networks in learning the latent complex interactions among the three elements of a triplet towards the neighbor node's confidence assignment.

Effect of Visual-Semantic Consistency Regularizer. In this part, we aim to examine the effect of the visual-semantic consistency regularizer on our model. Therefore, we removed the contrastive loss function from Eq. (5.16) by setting $\eta = 0$ and obtained a variant of MM-FRec, termed as MM-FRec-NoVSE. Table 5.5 shows the performance comparison between MM-FRec-NoVSE and MM-FRec. As can be seen, MM-FRec shows superiority over MM-FRec-NoVSE across all metrics. This verifies the benefit of bridging the semantic gap between the low-level visual clues and high-level attribute semantics.

Effect of Modality Fusion Manner. In previous experiments on model comparison, we have demonstrated the advantage of utilizing the multiple modalities of items in the fashion item recommendation. In this part, we aim to further explore the effect of the multi-modal fusion method. In particular, we introduced two variant models of MM-FRec, termed as MM-FRec$_+$ and MM-FRec$_{||}$, which adopt the addition and concatenation fusion strategies in Eq. (5.11), respectively. Table 5.6 shows the performance of our MM-FRec and its two variant models. It is obvious that our MM-FRec consistently outperforms the two variants over all evaluation metrics, implying that the element-wise product fusion strategy in MM-FRec is more powerful in our context of fashion item recommendation.

5.5.5 On Sensitivity Analysis

Besides, we studied the sensitivity of our model pertaining to the number of layers in the relation-specific neural networks, the number of propagation iterations, and the density of semantic attributes.

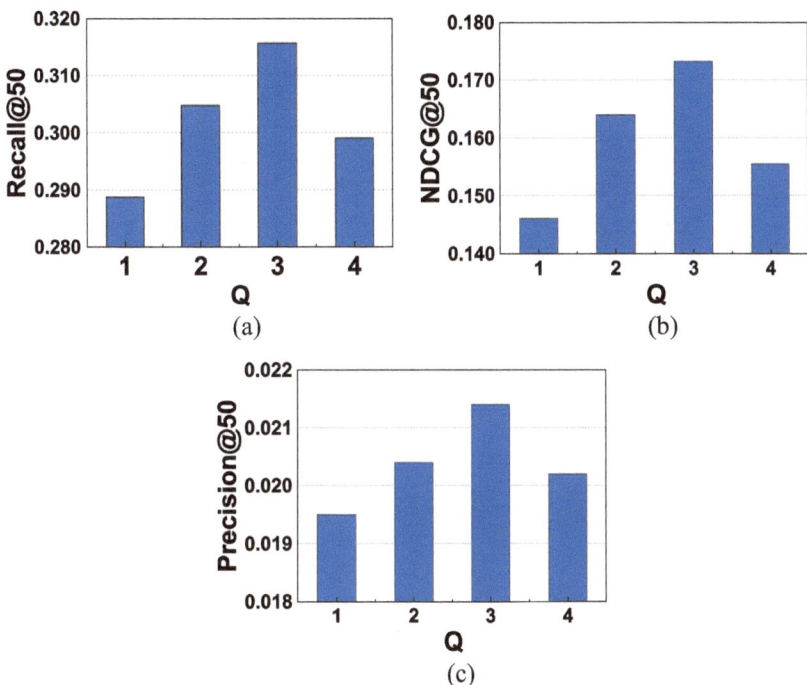

Fig. 5.4 Performance of our model with different numbers of layers in the relation-oriented network (i.e., Q)

On the Number of Layers in the Relation-specific Neural Networks. Firstly, we varied the number of layers in the relation-specific neural networks, i.e., Q, in the range of $\{1, 2, 3, 4\}$. Figure 5.4 shows the performance of our model with different Q in terms of Recall@50, NDCG@50, and Precision@50. As we can see, with the increase of the number of layers, the performance of our model increases first and then decreases. Overall, our model achieves the optimal performance when the number of layers in each relation-specific neural network is 3. When the number of layers is larger than 3, the model's performance begins to drop. The reason may be that too many layers may lead to the overfitting issue.

On the Number of Propagation Iterations. To learn the impact of the high-order connectivity, we changed the number of propagation iterations, i.e., L, from 1 to 4. Figure 5.5 illustrates the performance of our model with different L in terms of Recall@50, NDCG@50, and Precision@50. As we can see, similar to the previous experiment on Q, the performance of our model increases first and then decreases, with the increase of the number of propagation iterations. Our model performs best when $L = 3$. This demonstrates that the utility of the high-order connectivity information in the user and item latent representation learning. As for the performance drop of our model when $L = 4$, one likely reason is that with the richer information propagated with the increasing of L, certain noise can be also disseminated.

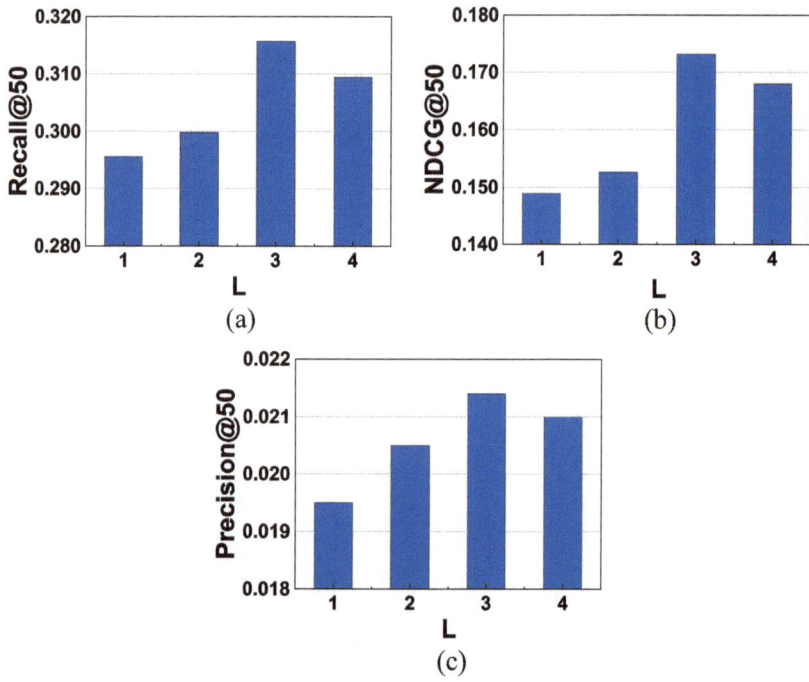

Fig. 5.5 Performance of our model with different numbers of propagation iterations (i.e., L)

On the Density of Semantic Attributes. As aforementioned, semantic attributes of items play an important role in characterizing the item features and benefiting the user preference modeling. Nevertheless, in real-world applications, the semantic attributes of items may not be rich enough. Therefore, we evaluated our model on different dataset configurations, which have different attribute densities. In other words, different proportions of semantic attribute labels (i.e., item-attribute edges) in our original dataset were made available. In particular, we adopted five different proportions: 0.2, 0.4, 0.6, 0.8, and 1.0. Figure 5.6 shows the performance of our model with different attribute density configurations in terms of Recall@50 and NDCG@50. As we can see, the more attribute tags we have, the better the performance of our model will be, which is in line with our expectation. In addition, although lacking attribute tags hurts the model performance, the degree of degradation is limited, reflecting a good tolerance of our model towards the density of semantic attributes. Notably, we also displayed the performance of the best baseline JNSKR with all attributes reserved by grey triangles in Fig. 5.6. As can be seen, our model with only 20% available semantic attributes can also surpass the best baseline that has 100% semantic attributes. This reconfirms the superiority of our model over existing methods.

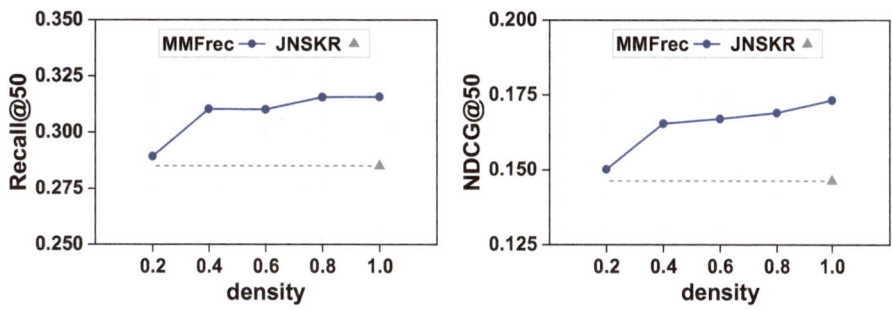

Fig. 5.6 Performance of our model with different dataset configurations, i.e., different densities of semantic attribute

5.5.6 On Case Study

To intuitively demonstrate the effectiveness of our model, apart from the quantitative evaluation, we also conducted the case study on both the item recommendation results and the recommendation interpretability.

To get the intuitive understanding of our MM-FRec in the task of item recommendation, we compared the recommendation results of MM-FRec and JNSKR for three testing users in Fig. 5.7. The reason why we chose JNSKR as the baseline lies in the fact that it is the best baseline for most metrics according to Table 5.3. Due to limited space, we list the top 10 recommended items' images of MM-FRec and JNSKR, respectively. To facilitate the understanding of the results, we also randomly sampled 10 item images that are historically

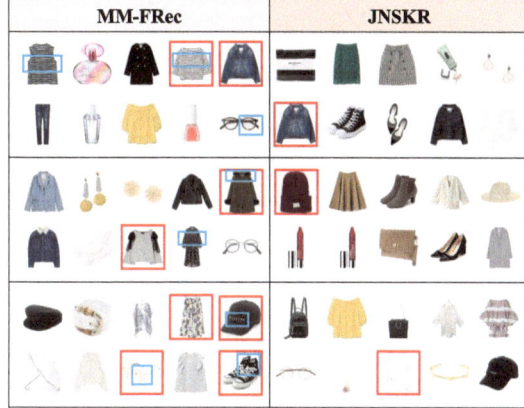

Fig. 5.7 Illustration of the returned Top-*10* recommendation results for three testing users of MM-FRec and JNSKR, where ground-truth images are highlighted by red boxes. Meanwhile, certain unique attributes captured by our model are also highlighted by blue boxes

interacted by these users in the training set to display the users' historical habits. Firstly, as can be seen, our model is able to return more ground-truth items that the user likes, highlighted in red boxes, than JNSKR, which reconfirms the superior recommendation performance of our model. Secondly, we noticed that our model is able to return items with unique attributes, highlighted in blue boxes, like the zebra pattern and round frame of glasses in the first case. One possible explanation for this observation is that MM-FRec copes well with the unique attributes, and hence promotes the recommendation performance. Similar conclusions can be found in the second and third cases.

To demonstrate the interpretability of our model, we randomly sampled a user-item pair, i.e., (u_8, p_{5472}), from the testing set, which is correctly predicted by our MM-FRec. We illustrated all accessible connections from u_8 to p_{5472} in the user-item-attribute tripartite graph $\mathcal{G}_a$ in Fig. 5.8. Notably, we omitted some weak connections according to the attention weights for clear illustration. As can be seen from the example, there are multiple paths linking the user u_8 and item p_{5472}, all of which indicate that u_8 has a tendency to like p_{5472}. After going through all paths, we noticed that the path $u_8 \xrightarrow{r_1} p_{15228} \xrightarrow{-r_1} u_{85} \xrightarrow{r_1} p_{5472}$, labeled by the blue solid line, gets the highest attention weight. This enables us to explain the recommendation result by the fact that the item p_{5472} has been interacted by user u_{85}, who once expressed the same interest in the item p_{15228} with the user u_8. The other connection paths in Fig. 5.8 can provide the supplement explanation for recommending item p_{5472} to

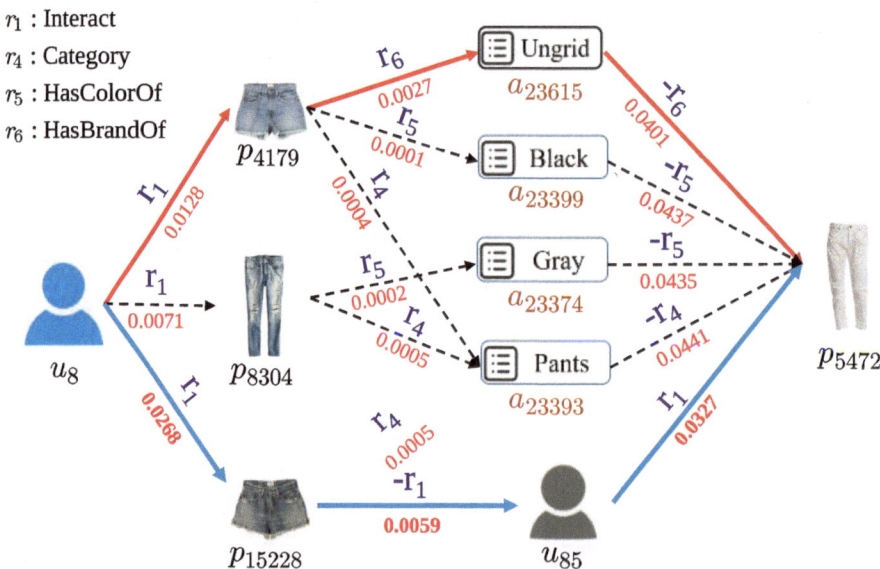

Fig. 5.8 Illustration of a recommendation result. All the paths are the accessible connections between the given user (i.e., u_8) and the recommended item (i.e., p_{5472}). The solid path in blue has the largest weight, while that in red has the second largest weight

user u_8. For example, based on the path $u_8 \xrightarrow{r_1} p_{4179} \xrightarrow{r_6} a_{23615} \xrightarrow{-r_6} p_{5472}$, labeled by the red solid line, we can conclude that one reason why user u_8 would be interested in item p_{5472} is that the item p_{5472} has the same brand (i.e., Ungrid) with the item p_{4179}, which is previously liked by user u_8.

5.6 Conclusion and Future Work

In this work, we present a Multi-Modal enhanced Fashion item Recommendation scheme, named MM-FRec, which jointly considers the visual images and semantic attributes of items to boost the recommendation performance. Meanwhile, we introduce a new relation-aware attention mechanism to adaptively measure how much information should be propagated from one's neighbor node according to not only the head and tail nodes but also the relation type. In particular, the multi-task learning framework is adopted to promote the triplet confidence assignment for unique attributes. Extensive experiments have been conducted on the real-world dataset and the results demonstrate the superiority of MM-FRec over existing methods and validate the benefits of jointly utilizing the attribute and vision modalities in the context of fashion item recommendation. In addition, experiments show that MM-FRec can not only provide explanations for the recommendation results according to the user-item-attribute tripartite graph, but also discover the user preference on unique attributes.

Currently, one limitation of our work is that we deem the user's preference temporally unchanged. However, in practice, the user's preference can be dynamically varied. Accordingly, in the future, we plan to incorporate the user's behavior trajectory to dynamically analyze user preferences for fashion items. Additionally, we are interested in integrating more clothing images of different views, such as front and back, and exploring more fine-grained visual preference mining methods to further improve the performance of fashion item recommendation.

References

1. Ai, Q., Azizi, V., Chen, X., Zhang, Y.: Learning heterogeneous knowledge base embeddings for explainable recommendation. Algorithms **11**(9), 137 (2018)
2. Bordes, A., Usunier, N., García-Durán, A., Weston, J., Yakhnenko, O.: Translating embeddings for modeling multi-relational data. In: Proceedings of the Advances in Neural Information Processing Systems, pp. 2787–2795 (2013)
3. Cao, D., Nie, L., He, X., Wei, X., Zhu, S., Chua, T.: Embedding factorization models for jointly recommending items and user generated lists. In: Proceedings of the International ACM SIGIR Conference on Research and Development in Information Retrieval, pp. 585–594. ACM (2017)
4. Cao, Y., Wang, X., He, X., Hu, Z., Chua, T.: Unifying knowledge graph learning and recommendation: Towards a better understanding of user preferences. In: Proceedings of the ACM International Conference on World Wide Web Conference, pp. 151–161. ACM (2019)

5. Chen, C., Zhang, M., Liu, Y., Ma, S.: Social attentional memory network: Modeling aspect- and friend-level differences in recommendation. In: Proceedings of the International Conference on Web Search and Data Mining, pp. 177–185. ACM (2019)

6. Chen, C., Zhang, M., Ma, W., Liu, Y., Ma, S.: Jointly non-sampling learning for knowledge graph enhanced recommendation. In: Proceedings of the International ACM SIGIR Conference on Research and Development in Information Retrieval, pp. 189–198. ACM (2020)

7. Chen, J., Wang, C., Wang, J.: Modeling the intransitive pairwise image preference from multiple angles. In: Proceedings of the ACM International Conference on Multimedia, pp. 351–359. ACM (2017)

8. Chen, W., Huang, P., Xu, J., Guo, X., Guo, C., Sun, F., Li, C., Pfadler, A., Zhao, H., Zhao, B.: Pog: personalized outfit generation for fashion recommendation at alibaba ifashion. In: Proceedings of the International ACM SIGKDD Conference on Knowledge Discovery and Data Mining, pp. 2662–2670 (2019)

9. Chen, X., Chen, H., Xu, H., Zhang, Y., Cao, Y., Qin, Z., Zha, H.: Personalized fashion recommendation with visual explanations based on multimodal attention network: Towards visually explainable recommendation. In: Proceedings of the International ACM SIGIR Conference on Research and Development in Information Retrieval, pp. 765–774 (2019)

10. Cheng, Z., Chang, X., Zhu, L., Kanjirathinkal, R.C., Kankanhalli, M.S.: MMALFM: explainable recommendation by leveraging reviews and images. ACM Transactions on Information Systems **37**(2), 16:1–16:28 (2019)

11. Glorot, X., Bengio, Y.: Understanding the difficulty of training deep feedforward neural networks. In: Proceedings of the International Conference on Artificial Intelligence and Statistics, vol. 9, pp. 249–256. JMLR.org (2010)

12. Gori, M., Pucci, A.: Itemrank: A random-walk based scoring algorithm for recommender engines. In: Proceedings of the International Joint Conference on Artificial Intelligence, pp. 2766–2771. IJCAI/AAAI Press (2007)

13. Gune, O., Banerjee, B., Chaudhuri, S., Cuzzolin, F.: Generalized zero-shot learning using generated proxy unseen samples and entropy separation. In: Proceedings of the ACM International Conference on Multimedia, pp. 4262–4270. ACM (2020)

14. Han, X., Song, X., Yin, J., Wang, Y., Nie, L.: Prototype-guided attribute-wise interpretable scheme for clothing matching. In: Proceedings of the International ACM SIGIR Conference on Research and Development in Information Retrieval, pp. 785–794 (2019)

15. He, K., Zhang, X., Ren, S., Sun, J.: Deep residual learning for image recognition. In: Proceedings of the IEEE/CVF Conference on Computer Vision and Pattern Recognition, pp. 770–778 (2016)

16. He, R., McAuley, J.: VBPR: visual bayesian personalized ranking from implicit feedback. In: Proceedings of the AAAI Conference on Artificial Intelligence, pp. 144–150. AAAI Press (2016)

17. He, X., Deng, K., Wang, X., Li, Y., Zhang, Y., Wang, M.: Lightgcn: Simplifying and powering graph convolution network for recommendation. In: Proceedings of the International ACM SIGIR Conference on Research and Development in Information Retrieval, pp. 639–648. ACM (2020)

18. He, X., Liao, L., Zhang, H., Nie, L., Hu, X., Chua, T.: Neural collaborative filtering. In: Proceedings of the ACM International Conference on World Wide Web, pp. 173–182. ACM (2017)

19. Hou, M., Wu, L., Chen, E., Li, Z., Zheng, V.W., Liu, Q.: Explainable fashion recommendation: A semantic attribute region guided approach. In: Proceedings of the International Joint Conference on Artificial Intelligence, pp. 4681–4688 (2019)

20. Huang, X., Fang, Q., Qian, S., Sang, J., Li, Y., Xu, C.: Explainable interaction-driven user modeling over knowledge graph for sequential recommendation. In: Proceedings of the ACM International Conference on Multimedia, pp. 548–556. ACM (2019)

21. Järvelin, K., Kekäläinen, J.: IR evaluation methods for retrieving highly relevant documents. In: Proceedings of the International ACM SIGIR Conference on Research and Development in Information Retrieval, pp. 41–48. ACM (2000)

22. Kingma, D.P., Ba, J.: Adam: A method for stochastic optimization. In: Proceedings of the International Conference on Learning Representations, pp. 1–15 (2014)
23. Lin, Y., Liu, Z., Sun, M., Liu, Y., Zhu, X.: Learning entity and relation embeddings for knowledge graph completion. In: Proceedings of the International Joint Conference on Artificial Intelligence, pp. 2181–2187. AAAI Press (2015)
24. Liu, W., Zhang, C., Lin, G., Hung, T., Miao, C.: Weakly supervised segmentation with maximum bipartite graph matching. In: Proceedings of the ACM International Conference on Multimedia, pp. 2085–2094. ACM (2020)
25. Liu, Y., Yang, S., Lei, C., Wang, G., Tang, H., Zhang, J., Sun, A., Miao, C.: Pre-training graph transformer with multimodal side information for recommendation. In: Proceedings of the ACM International Conference on Multimedia, pp. 2853–2861. ACM (2021)
26. McAuley, J., Targett, C., Shi, Q., Van Den Hengel, A.: Image-based recommendations on styles and substitutes. In: Proceedings of the International ACM SIGIR Conference on Research and Development in Information Retrieval, pp. 43–52 (2015)
27. Pu, N., Chen, W., Liu, Y., Bakker, E.M., Lew, M.S.: Dual gaussian-based variational subspace disentanglement for visible-infrared person re-identification. In: Proceedings of the ACM International Conference on Multimedia, pp. 2149–2158. ACM (2020)
28. Qiu, J., Tang, J., Ma, H., Dong, Y., Wang, K., Tang, J.: Deepinf: Social influence prediction with deep learning. In: Proceedings of the International ACM SIGKDD Conference on Knowledge Discovery and Data Mining, pp. 2110–2119. ACM (2018)
29. Rendle, S., Freudenthaler, C., Gantner, Z., Schmidt-Thieme, L.: BPR: bayesian personalized ranking from implicit feedback. In: Proceedings of the Conference on Uncertainty in Artificial Intelligence, pp. 452–461. AUAI Press (2009)
30. Song, X., Han, X., Li, Y., Chen, J., Xu, X.S., Nie, L.: Gp-bpr: Personalized compatibility modeling for clothing matching. In: Proceedings of the ACM International Conference on Multimedia, pp. 320–328 (2019)
31. Sun, L., Lian, Z., Tao, J., Liu, B., Niu, M.: Multi-modal continuous dimensional emotion recognition using recurrent neural network and self-attention mechanism. In: Proceedings of the ACM International Conference on Multimedia, pp. 27–34. ACM (2020)
32. Velickovic, P., Cucurull, G., Casanova, A., Romero, A., Liò, P., Bengio, Y.: Graph attention networks. In: Proceedings of the International Conference on Learning Representations, pp. 1–12. OpenReview.net (2018)
33. Wang, H., Zhao, M., Xie, X., Li, W., Guo, M.: Knowledge graph convolutional networks for recommender systems. In: Proceedings of the ACM International Conference on World Wide Web Conference, pp. 3307–3313. ACM (2019)
34. Wang, X., He, X., Cao, Y., Liu, M., Chua, T.: KGAT: knowledge graph attention network for recommendation. In: Proceedings of the International ACM SIGKDD Conference on Knowledge Discovery and Data Mining, pp. 950–958. ACM (2019)
35. Wang, X., He, X., Wang, M., Feng, F., Chua, T.S.: Neural graph collaborative filtering. In: Proceedings of the International ACM SIGIR Conference on Research and Development in Information Retrieval, pp. 165–174. ACM (2019)
36. Xiangnan, H., Ming, G., Min-Yen, K., Dingxian, W.: Birank: towards ranking on bipartite graphs. IEEE Transactions on Knowledge and Data Engineering 29(1), 57–71 (2016)
37. Xu, C., Guan, Z., Zhao, W., Wu, Q., Yan, M., Chen, L., Miao, Q.: Recommendation by users' multimodal preferences for smart city applications. IEEE Transactions on Industrial Informatics 17(6), 4197–4205 (2021)
38. Yang, X., Song, X., Han, X., Wen, H., Nie, J., Nie, L.: Generative attribute manipulation scheme for flexible fashion search. In: Proceedings of the International ACM SIGIR Conference on Research and Development in Information Retrieval, pp. 941–950. ACM (2020)

39. Yu, W., Zhang, H., He, X., Chen, X., Xiong, L., Qin, Z.: Aesthetic-based clothing recommendation. In: Proceedings of the ACM International Conference on World Wide Web Conference, pp. 649–658. ACM (2018)
40. Zhang, F., Yuan, N.J., Lian, D., Xie, X., Ma, W.: Collaborative knowledge base embedding for recommender systems. In: Proceedings of the International ACM SIGKDD Conference on Knowledge Discovery and Data Mining, pp. 353–362. ACM (2016)
41. Zhou, Z., Di, X., Zhou, W., Zhang, L.: Fashion sensitive clothing recommendation using hierarchical collocation model. In: Proceedings of the ACM International Conference on Multimedia, pp. 1119–1127. ACM (2018)

Hashing-Based Efficient Outfit Recommendation

6

6.1 Introduction

In this chapter, we explore the use of hashing-based methods for efficient outfit recommendation. The recent surge in fashion-oriented products on e-commerce platforms has made it increasingly challenging and time-consuming for users to find their desired items among the numerous available options. In the light of this, fashion recommendation has gained increasing attention from both industry and academia, due to its great economic value and benefit in improving the users' consumption experience. Similar to the traditional news or movie recommendation, most of previous studies [21, 33] focus on recommending a single fashion item to the given user. Nevertheless, what the user really needs in practice is usually a well-matched outfit (i.e., a set of compatible and complementary fashion items), rather than a single fashion item. Inspired by this, a few efforts have been gradually shifted to study Personalized Outfit Recommendation (POR). For example, Li et al. [15] addressed this task by adopting Graph Neural Networks (GNNs) [6] to learn outfit and user representations. Lin et al. [16] proposed the OutfitNet model via leveraging attention-based multiple instance learning. Despite their promising performance, they merely focus on improving the recommendation effectiveness, while ignoring the efficiency. In fact, the number of available outfit candidates on fashion platforms has kept increasing exponentially, which implies the necessity of deploying an efficient POR system. Towards this end, Lu et al. [20] proposed to learn a hash binary code for each fashion entity (i.e., the user and item). Nevertheless, they overlooked the influence of attributes for POR. In fact, attributes usually characterize key semantic information of an item, which can be hardly expressed by the visual image, such as the material and price. Therefore, in this work, we bring in attributes associated with fashion items, and work towards fully exploring all entities (i.e., users, outfits, items and attributes) and their various relations (i.e., user-outfit, outfit-item, and item-attribute interactions) to further promote the performance of the efficient POR.

© The Author(s), under exclusive license to Springer Nature Switzerland AG 2025
W. Guan et al., *Advanced Multimodal Compatibility Modeling and Recommendation*,
Synthesis Lectures on Information Concepts, Retrieval, and Services,
https://doi.org/10.1007/978-3-031-81048-0_6

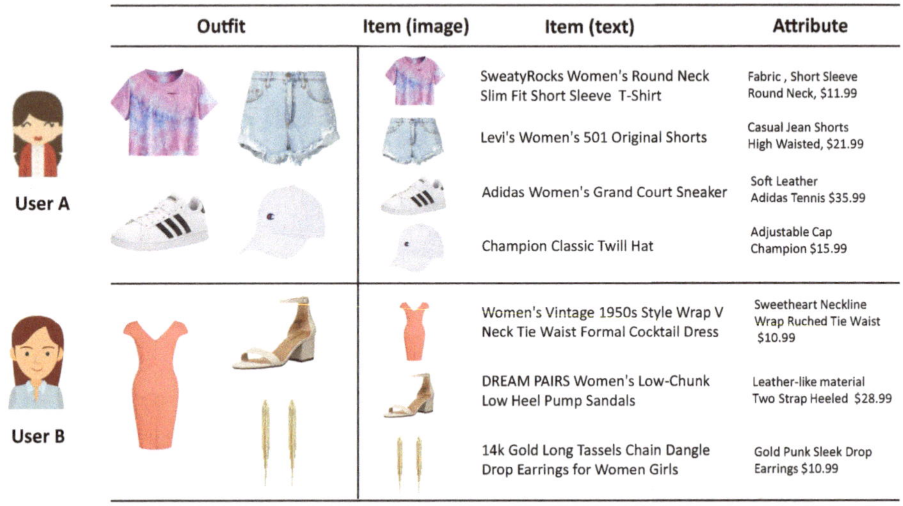

	Outfit	Item (image)	Item (text)	Attribute
User A			SweatyRocks Women's Round Neck Slim Fit Short Sleeve T-Shirt	Fabric , Short Sleeve Round Neck, $11.99
			Levi's Women's 501 Original Shorts	Casual Jean Shorts High Waisted, $21.99
			Adidas Women's Grand Court Sneaker	Soft Leather Adidas Tennis $35.99
			Champion Classic Twill Hat	Adjustable Cap Champion $15.99
User B			Women's Vintage 1950s Style Wrap V Neck Tie Waist Formal Cocktail Dress	Sweetheart Neckline Wrap Ruched Tie Waist $10.99
			DREAM PAIRS Women's Low-Chunk Low Heel Pump Sandals	Leather-like material Two Strap Heeled $28.99
			14k Gold Long Tassels Chain Dangle Drop Earrings for Women Girls	Gold Punk Sleek Drop Earrings $10.99

Fig. 6.1 Examples of users' outfit compositions

Inspired by the remarkable success of graph learning in entity representation learning [28] and outfit recommendation [15], we resort to the graph learning technique to fulfill the efficient POR. However, this is non-trivial due to the following three key challenges. **C1: Heterogeneous fashion entities.** As shown in Fig. 6.1, our work involves four kinds of entities, namely, users, outfits, items and attributes, where some entities (e.g., items, and attributes) have explicit contents, while others (e.g., users and outfits) have not. Therefore, how to seamlessly unify these heterogeneous entities within a graph is the tough challenge we face. **C2: Efficient graph convolution.** Traditional graph convolution methods focus on propagating message among graph nodes and updating entity representations simultaneously at the end of the whole graph convolution. Such graph convolution schemes suffer from two key limitations. First, they do not explicitly distinguish the types of entities during information propagation. Second, their convolution is not totally efficient, especially for our heterogeneous graph that involves multiple types of entities and relations. This is because once a type of entity representations get updated, they can be used timely for updating other types of entities. Therefore, how to fulfill the efficient graph convolution is another challenge. **C3: Information preservation during hashing.** Converting the continuous hash representation of each entity to the binary hash code inevitably loses certain information. Therefore, how to retain the original information to the greatest extent during the hash code learning constitutes the third challenge.

To address these challenges, we first organize the user, outfit, item and attribute entities in the context of POR into a unified heterogeneous four-partite graph. Specifically, the nodes of this graph correspond the four types of entities, and are linked by three types of edges: user-outfit interactions, outfit-item relations, and item-attribute association relations. We

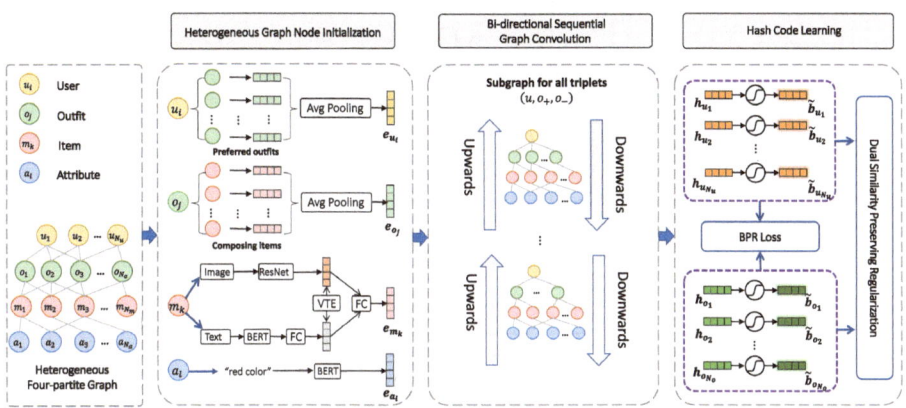

Fig. 6.2 Illustration of the proposed BiHGH scheme. It consists of three key components: (1) heterogeneous graph node initialization, (2) bi-directional sequential graph convolution, and (3) hash code learning

then devise a novel bi-directional heterogeneous graph hashing scheme, called BiHGH, as shown in Fig. 6.2. This scheme consists of three key components: heterogeneous graph node initialization, bi-directional sequential graph convolution, and hash code learning. The first component works on embedding entities in our heterogeneous graph from the content level. It is worth mentioning that instead of using the conventional fixed one-hot or learn-from-scratch embedding, we initialize each outfit entity based upon the contents of its composing item entities, and each user entity by its historically interacted outfit entity embeddings. Moreover, we incorporate the contrastive loss [5] to encourage the semantic consistency between visual and textual modalities. In the second component, we first divide the four-partite graph into three subsequent subgraphs, namely user-outfit subgraph, outfit-item subgraph, and item-attribute subgraph, as illustrated in Fig. 6.3. We then conduct bi-directional graph convolution over the subgraphs sequentially and iteratively upwards and downwards. In this manner, we can separately deal with different types of neighbor entities and reduce the computational cost of the big graph convolution. As to the third one, it aims to obtain the binary hash code of each entity. To accomplish this, we adopt the Bayesian Personalized Ranking (BPR) [23] loss for optimization, which has been proven to be effective in the user preference modeling. Meanwhile, to enhance the hash code learning of users and outfits, we design a dual similarity preserving regularization to remain information during hashing. The underlying philosophy is that if two users prefer the same outfit or two outfits share a certain item, their corresponding hash codes should be similar. We conduct extensive experiments on the benchmark dataset, and the results demonstrate the superiority of our model to several cutting-edge baselines. We have released our code and parameters.[1]

---

[1] https://outfitrec.wixsite.com/bihgh.

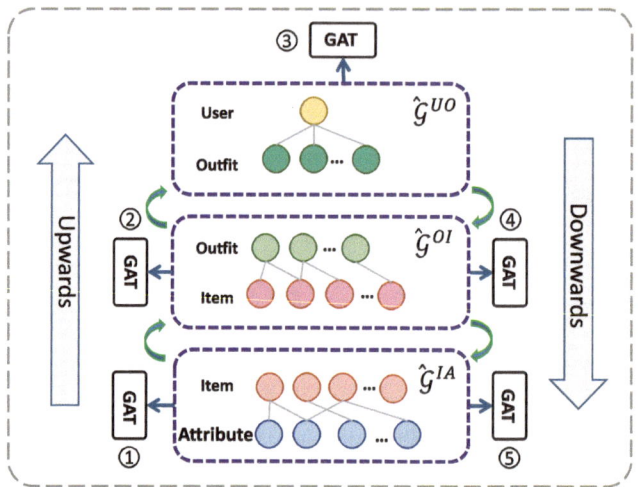

Fig. 6.3 Illustration of the proposed bi-directional sequential graph convolution algorithm

Our main contributions can be highlighted in threefold:

1. We present a novel heterogeneous graph learning-based outfit recommendation scheme, where the four types of entities (i.e., users, outfits, items, and attributes) and their relations are seamlessly integrated.
2. We creatively devise a bi-directional graph convolution algorithm to fulfill the efficient optimization, which works on sequentially transferring knowledge via repeating upwards and downwards optimization. It greatly alleviates the computational cost of big graph convolution.
3. We design a dual (i.e., outfit-level and user-level) similarity preserving regularizations to prevent the information loss during the hash learning.

6.2   Related Work

6.2.1   Fashion Recommendation

Existing studies on fashion recommendation [30, 32, 34] can be roughly divided into two groups: item and outfit recommendation.

Item recommendation focuses on recommending a single item to the given user. For example, He et al. [11] developed a fashion item recommendation model based on the matrix factorization framework [14], which incorporates visual information of products to enhance the item representation. Later, Yu et al. [33] introduced a brain-inspired deep network to promote user preference modeling from the aesthetic perspective for fashion item recommendation. Towards explainable recommendation, Hou et al. [12] proposed a semantic attribute region guided approach, which extracts attributes from the item image, and employs the attention mechanism to explain the item recommendation result. Differently, Song et al. [25] noticed the necessity of considering the items that the user already has when recommending items, and hence investigated the task of recommending a complementary and compatible item (e.g., bottom) for a given user with a given item (e.g., top). In particular, they proposed a personalized compatibility modeling scheme for clothing matching, whereby both the user-item preference and the item-item compatibility are jointly modeled for personalized complementary item recommendation.

By contrast, the outfit recommendation aims to recommend the whole outfits, i.e., the sets of complementary items, to match the preference of a given user. For instance, Chen et al. [2] proposed a personalized outfit generation model that is able to generate outfits for the particular user by jointly considering the images and textual descriptions of items. Instead of generating outfits, Li et al. [15] proposed a hierarchical fashion graph network for POR, which models the relationships among users, items, and outfits, simultaneously. Meanwhile, Lin et al. [16] developed a bi-stage outfit recommendation system. It learns the compatibility between fashion items in the first stage, and then predicts the users' preference to outfits based on the visual information of items in the second one.

Although these studies have achieved great progress, they focus on exploring the user, outfit, and item entities. In other words, they overlook the semantic attributes of items, which actually well characterize the key features of items and reflect the user's specific preferences to items from the semantic perspective. In addition, existing methods generally focus on improving the recommendation effectiveness, while ignoring the recommendation efficiency. In light of this, we incorporated the attribute entities to promote the POR effectiveness, and explored the hashing methods to improve the model efficiency.

6.2.2   Learning to Hash

Learning to hash, which aims to learn a compact binary hash code for the given instance, has attracted a large amount of research interest [24, 27], due to its prominent advantages in saving the time and storage costs. One key feature of hashing methods [17, 18, 39] is that the $sign$ function is usually used to convert the continuous representations of instances into binary hash codes. This hinders the direct optimization of the discrete hash codes. Therefore, most hashing methods would allow some relaxations over the optimization, that is, optimizing the model over the continuous item hash representation, which can be seen as the surrogates of the binary hash codes. To reduce the quantization error, HashNet [1] leverages the scaled $tanh$ function to approximate the $sign$ function.

Inspired by its astonishing success in efficiency improvement, hash learning has been adopted in several recommendation tasks [20, 37]. For example, Zhou et al. [37] investigated the efficient item recommendation by learning binary codes for collaborative filtering. The hamming distance between the binary codes of the user and the item is used for measuring the user's preference over items. Lu et al. [20] proposed to learn binary codes for fashion entities towards efficient personalized outfit recommendation, where the BPR loss for user preference modeling and contrastive loss for vision-text embedding regularization are used for optimization.

Despite their significance in efficient recommendation, they cannot effectively prevent the information loss during hash code learning. Towards this end, in this work, we designed a dual similarity preserving regularization and hence strengthen the discriminative binary representation of various fashion entities.

6.3 Methodology

In this section, we first formulate the research problem, and then detail the three components of our proposed BiHGH scheme.

6.3.1 Problem Formulation

In this work, we focus on the task of POR, which targets at recommending an outfit rather than a single fashion item to a given user. Suppose that we have a set of N_u users $\mathcal{U} = \{u_1, u_2, \ldots, u_{N_u}\}$, a set of N_o outfits $O = \{o_1, o_2, \ldots, o_{N_o}\}$, a set of N_m items $M = \{m_1, m_2, \ldots, m_{N_m}\}$, and a set of N_a attributes $\mathcal{A} = \{a_1, a_2, \ldots, a_{N_a}\}$. Each user u_i historically interacts with a set of preferred outfits $O_i \subseteq O$, each outfit o_j is composed by a set of compatible items, denoted as $M_j \subseteq M$, and each item m_k is associated with a set of attributes $\mathcal{A}_k \subseteq \mathcal{A}$. Moreover, as illustrated in Fig. 6.1, each item m_j also involves an image v_j and a textual description t_j.

We resort to a heterogeneous four-partite graph to organize entities within a unified structure. In particular, we denote the graph as $G = (\mathcal{E}, \mathcal{R})$, where $\mathcal{E} = \mathcal{U} \cup O \cup M \cup \mathcal{A}$ represents the set of entity nodes, and $\mathcal{R}$ denotes the set of edges linking these nodes. It is noteworthy that edges characterize various relations among entities, including user-outfit historical interactions, outfit-item subordinating relations, and item-attribute association relations. In addition, we have a set of triplet training samples $\mathcal{D} = \{(u_i, o_j, o_k) | u_i \in \mathcal{U}, o_j \in O, o_k \in O, (u_i, o_j) \notin \mathcal{E}\}$, where o_j and o_k denote the positive and negative outfit according to the user u_i's preference, respectively. In a sense, based on the heterogenerous four-partite graph and training set $\mathcal{D}$, we aim to optimize a model $\mathcal{F}$, which can accurately and efficiently estimate the preference degree of a given user towards an arbitrary outfit as follows,

$$s_{ij} = \mathcal{F}(u_i, o_j | \Theta_F, G), \tag{6.1}$$

where s_{ij} denotes the preference of the user u_i to the outfit o_j, and Θ_F refers to the to-be-learned model parameters.

6.3.2 BiHGH

In this subsection, we elaborate the proposed BiHGH.

Heterogeneous Graph Node Initialization. This component aims to initialize node-level embeddings in the heterogeneous graph. As the heterogeneous graph has four types of entities, we introduce their corresponding embedding methods one by one.

Item Embedding. Inspired by previous studies [7, 9, 36], we jointly exploit the image and textual information to learn the item embedding. In particular, we utilize the pre-trained ResNet [8, 10], which will be fine-tuned during training, to obtain the item's visual feature. Meanwhile, we adopt the pre-trained BERT [4] as well as a fully-connected layer for fine-tuning, to obtain the item's textual feature. More concretely, we average the hidden states corresponding to the special token [CLS] of the last two layers as the textual feature. We ultimately concatenate the visual and textual features of each item, and feed it into a learnable fully-connected layer to derive the final item embedding. Formally, we have

$$
\begin{cases}
\mathbf{e}_{v_k} = \text{ResNet}\,(v_k)\,, \\
\mathbf{e}_{t_k} = f_t(\text{Bert}\,(t_k)_{[CLS]}), \\
\mathbf{e}_{m_k} = f_m\left(\mathbf{e}_{v_k}\|\mathbf{e}_{t_k}\right),
\end{cases}
\tag{6.2}
$$

where f_t and f_m denote the learnable fully-connected layers. $\mathbf{e}_{v_k} \in \mathbb{R}^d$ and $\mathbf{e}_{t_k} \in \mathbb{R}^d$ are the d-dimensional visual and textual embedding of the item m_k, respectively. $\|$ is the concatenation operation, and $\mathbf{e}_{m_k} \in \mathbb{R}^d$ is the final initial embedding of the item m_k.

To regularize the semantic consistency between visual and textual embeddings of the same item, we adopt the following contrastive loss,

$$
\begin{aligned}
\mathcal{L}_{VTE} = &\sum_{\substack{i \neq k}}^{N_m} max\{0, c - d(\mathbf{e}_{v_i}, \mathbf{e}_{t_i}) + d(\mathbf{e}_{v_i}, \mathbf{e}_{t_k})\} \\
+ &\sum_{\substack{i \neq k}}^{N_m} max\{0, c - d(\mathbf{e}_{v_i}, \mathbf{e}_{t_i}) + d(\mathbf{e}_{v_k}, \mathbf{e}_{t_i})\},
\end{aligned}
\tag{6.3}
$$

where c is a margin hyperparameter and $d(\mathbf{a}, \mathbf{b}) = \mathbf{a}^T\mathbf{b}$.

Attribute Embedding. To mine the semantic content of each attribute, we also resort to the pre-trained BERT model with a learnable fully-connected layer, to derive its embedding. For each attribute entity a_l, its embedding can be defined as follows,

$$
\mathbf{e}_{a_l} = f_a(\text{Bert}\,(a_l)_{[CLS]}),
\tag{6.4}
$$

where $\mathbf{e}_{a_l} \in \mathbb{R}^d$ stands for the initial embedding of the attribute entity a_l, and f_a denotes the fully-connected layer to fine-tune the embedding. Notably, considering that attributes are very short compared with the textual descriptions, we only adopt the representation of the special token [CLS] from the last layer of BERT as the attribute embedding.

Outfit Embedding. Each outfit is a set of fashion items and hence has no concrete content information. We thus derive the initial embedding of each outfit by aggregating all the embeddings of its composing items as,

$$\mathbf{e}_{o_j} = \mathrm{Avg}(m_k | m_k \in \mathcal{M}_j), \tag{6.5}$$

where $\mathbf{e}_{o_j} \in \mathbb{R}^d$ denotes the initial embedding of the outfit o_j. $\mathrm{Avg}(\cdot)$ refers to the average pooling operation.

User Embedding. Analogous to the outfit, each user entity also has no straightforward content information. We hence fuse all the embeddings of his/her historically interacted and preferred outfits, to derive the user embedding as follows,

$$\mathbf{e}_{u_i} = \mathrm{Avg}(o_j | o_j \in \mathcal{O}_i), \tag{6.6}$$

where $\mathbf{e}_{u_i} \in \mathbb{R}^d$ denotes the initial embedding of the user u_i.

Bi-directional Sequential Graph Convolution. To perform the heterogeneous graph learning and refine the embedding of each entity, we build a specific graph $\hat{\mathcal{G}}$ for each training triplet sample (u, o_+, o_-) according to the whole graph $\mathcal{G}$. In particular, we here temporally treat the whole graph $\mathcal{G}$ as a tree structure, whereby the user, outfit, item, and attribute entities are positioned in the 1st, 2nd, 3rd, and 4th levels, respectively, as shown in Fig. 6.3. Then, the entity set of $\hat{\mathcal{G}}$ includes all the children entity nodes of these three entity nodes, i.e., u, o_+, and o_-. We thereafter gather all the edges among these entities in the graph $\mathcal{G}$ as the edge set of $\hat{\mathcal{G}}$. To perform graph learning, the most straightforward way is to conduct the conventional graph convolution operations, like GAT, which refine entity embeddings by simultaneously propagating information among entity nodes. Although feasible, we argue that this manner suffers from two key limitations. First, it fails to distinguish the types of neighbor nodes during information propagation. Second, as this graph has four kinds of entities, the graph convolution requires extremely powerful servers and its computational cost is relatively high, according to our preliminary attempts.

To address these issues, we propose to conduct the bi-directional sequential graph convolution to refine each node embedding. In particular, we decompose the entire graph into three subgraphs: user-outfit graph, outfit-item graph, and item-attribute graph. They are respectively denoted as $\hat{\mathcal{G}}^{UO}$, $\hat{\mathcal{G}}^{OI}$, and $\hat{\mathcal{G}}^{IA}$. Following that, we conduct graph convolution in two directions: upwards (i.e., in the order of $\hat{\mathcal{G}}^{IA}, \hat{\mathcal{G}}^{OI}, \hat{\mathcal{G}}^{UO}$) and then downwards (i.e., in the order of $\hat{\mathcal{G}}^{UO}, \hat{\mathcal{G}}^{OI}, \hat{\mathcal{G}}^{IA}$), as illustrated in Fig. 6.3. In other words, in each training iteration, we conduct five subgraph convolution sequentially over the subgraphs $\hat{\mathcal{G}}^{IA}, \hat{\mathcal{G}}^{OI}, \hat{\mathcal{G}}^{UO}, \hat{\mathcal{G}}^{OI}, \hat{\mathcal{G}}^{IA}$. It is worth noting that once the entity embedding is updated by one subgraph convolution, it is then directly used at the next one, rather than waiting for the end of all subgraph learning. By doing so, we can timely utilize the messages passed by various entities and promote the model convergence.

In this work, we adopt the same graph convolution module for all subgraphs and both directions. We here take the graph convolution of the subgraph $\hat{\mathcal{G}}^{OI}$ as an example, and the rest follows the same procedure. Suppose that our graph convolution module has L layers. According to GAT [21], the l-th layer of graph convolution towards the outfit entity o in the subgraph $\hat{\mathcal{G}}^{OI}$ can be written as,

$$
\begin{cases}
\mathbf{e}_o^{l+1} = \mathbf{e}_o^l + \|_{k=1}^M \left(\phi \left(\sum_{m_s \in \mathcal{N}_o} \alpha_{m_s}^{lk} \mathbf{W}_k^l \mathbf{e}_{m_s}^l \right) \right), l = 0, \ldots, L-1, \\
\alpha_{m_s}^{lk} = softmax(\eta(\boldsymbol{\gamma}_l^T [\mathbf{W}_k^l \mathbf{e}_o^l \| \mathbf{W}_k^l \mathbf{e}_{m_s}^l])),
\end{cases}
\tag{6.7}
$$

where $\mathcal{N}_o$ denotes the set of one-hop neighbors of the outfit entity o. The symbols $\mathbf{e}_o^l$ and $\mathbf{e}_{m_s}^l$ refer to the embedding of the outfit o and the item m_s derived by the l-th layer, respectively. $\boldsymbol{\gamma}_l$ and $\mathbf{W}_k^l$ are the to-be-learned parameters. M is the number of heads in the multi-head attention. η and ϕ are the LeakyReLU and Exponential Linear Unit (ELU) activation functions, respectively. The graph convolution towards each item entity m_s in the subgraph $\hat{\mathcal{G}}^{OI}$ can be defined in a similar way. Notably, $\mathbf{e}_o^0$ and $\mathbf{e}_{m_s}^0$ are the initial representations of the outfit entity o and item entity m_s for the convolution over the subgraph $\hat{\mathcal{G}}^{OI}$, respectively, which are different for different directions' subgraph convolution. In the upwards direction, they are yielded by the graph convolution over the subgraph $\hat{\mathcal{G}}^{IA}$. While in the downwards one, they are produced by that over the subgraph $\hat{\mathcal{G}}^{UO}$.

We denote the refined user, outfit, item, and attribute embeddings as $\tilde{\mathbf{e}}_u$, $\tilde{\mathbf{e}}_o$, $\tilde{\mathbf{e}}_m$, and $\tilde{\mathbf{e}}_a$, respectively. We ultimately adopt the MLP to derive the final hash representation for each entity as $\mathbf{h}_x = MLP(\tilde{\mathbf{e}}_x)$, where $x = \{u, o, m, a\}$, and $\mathbf{h}_x \in \mathbb{R}^D$.

Hash Code Learning. To improve the efficiency of POR, we introduce the hash layer to learn the binary codes for user and outfit entities. Due to the non-differentiable problem over discrete hash functions, we adopt the $tanh$ function to approximate the $sign$ operation and derive the continuous hash codes as,

$$
\begin{cases}
\tilde{\mathbf{b}}_{u_i} = tanh(\beta \mathbf{h}_{u_i}), \\
\tilde{\mathbf{b}}_{o_j} = tanh(\beta \mathbf{h}_{o_i}),
\end{cases}
\tag{6.8}
$$

where β is a scale parameter to ensure the suitable interval before $tanh$ function. $\tilde{\mathbf{b}}_{u_i}$ and $\tilde{\mathbf{b}}_{o_j}$ refer to the continuous hash codes of the user u_i and outfit o_j, respectively.

6.3.3   Optimization

Towards optimization, we adopt the conventional BPR loss for user preference learning and design the dual similarity preserving regularization to enhance the hash code learning.

BPR Loss for User Preference Learning. Based on our training set of triplets, i.e., $\mathcal{D}$, the BPR loss can be formulated as follows,

$$\begin{cases} \mathcal{L}_{BPR} = - \displaystyle\sum_{(u_i,o_j,o_k)\in\mathcal{D}} log(1 + exp(s_{ij} - s_{ik})), \\ \\ s_{ij} = \tilde{\mathbf{b}}_{u_i}^T \tilde{\mathbf{b}}_{o_j}, \end{cases} \tag{6.9}$$

where s_{ij} stands for the preference of the user u_i towards the outfit o_j, calculated by the inner product of the continuous hash codes of the user u_i and outfit o_j. Intuitively, we expect that the user's preference towards the positive outfit o_j should be larger than that to the negative one o_k.

Dual Similarity Preserving Regularization. Similarity preserving regularization has proven to be effective in supervising the hash code learning by remaining the original pairwise semantic similarity [3]. As each outfit is composed by a set of complementary items, these composing items can be seen as labels of the outfit entity. Similarly, each user prefers a few outfits, and thus each user entity can be analogously labeled by outfits. Based on these labels, we can derive the ground truth similarity matrix for the user and outfit entities. Formally, let $\mathbf{S}^o \in \mathbb{R}^{N_o \times N_o}$ and $\mathbf{S}^u \in \mathbb{R}^{N_u \times N_u}$ be the ground truth similarity matrices for outfit and user entities, respectively. In particular, the (i, j)-th entry $S_{ij}^o = 1$, if the outfits o_i and o_j share at least one composing item, otherwise $S_{ij}^o = 0$. Similarly, the (i, j)-th entry $S_{ij}^u = 1$, if both users u_i and u_j prefer the same outfit, otherwise $S_{ij}^u = 0$.

Afterwards, we deploy the dual similarity preserving regularization, with the goal of maximizing the Hamming distance between two entities whose semantic similarity is 0, while minimizing those with semantic similarity as 1. To be specific, inspired by [26, 29], we define the following regularizations,

$$\begin{cases} \mathcal{L}_{S_o} = - \displaystyle\sum_{i,j=1}^{N_o} \left(S_{ij}^o \phi_{ij}^o - log(1 + e^{\phi_{ij}^o}) \right), \quad \phi_{ij}^o = cos(\tilde{\mathbf{b}}_{o_i}, \tilde{\mathbf{b}}_{o_j}), \\ \\ \mathcal{L}_{S_u} = - \displaystyle\sum_{i,j=1}^{N_u} \left(S_{ij}^u \phi_{ij}^u - log(1 + e^{\phi_{ij}^u}) \right), \quad \phi_{ij}^u = cos(\tilde{\mathbf{b}}_{u_i}, \tilde{\mathbf{b}}_{u_j}), \end{cases} \tag{6.10}$$

where ϕ_{ij}^o refers to the hash code based semantic similarity between outfits o_i and o_j, ϕ_{ij}^u is the one between users u_i and u_j, and $cos(\cdot, \cdot)$ denotes the cosine function. In a sense, we expect that the hash representation based semantic similarities approximate the ground truth ones.

Ultimately, we reach the final objective function as follows,

$$\mathcal{L} = \mathcal{L}_{BPR} + \alpha_1 \mathcal{L}_{VTE} + \alpha_2(\gamma \mathcal{L}_{S_o} + \mathcal{L}_{S_u}), \tag{6.11}$$

where α_1, α_2, and γ are the non-negative hyperparamters controlling the importance of the corresponding regularizations.

6.4 Experiments

In this section, we conducted experiments over the real-world dataset by answering the following research questions.

1. Does BiHGH outperform state-of-the-art baselines?
2. How does each component affect BiHGH?
3. How sensitive is BiHGH?
4. What is the intuitive performance of BiHGH?

6.4.1 Experimental Settings

Dataset. In general, there are three popular datasets for personalized outfit recommendation: Polyvore-630 [20], Polyvore-519 [20], and IQON-550 [19]. However, the former two datasets lack the attribute information of items. Therefore, we adopted the dataset IQON-550 that has rich attribute labels to evaluate our method in the context of POR. Derived from the dataset IQON3000 [25], this dataset consists of 550 users with 57,750 positive outfits, where each user historically interacted with 120 positive outfits. For each user, the 120 positive outfits are divided into three parts: 85, 15, and 20 outfits for training, validation, and testing, respectively. It is worth noting that we further split the training set into 2 chunks: 50% for building the whole heterogeneous graph G, and 50% for creating the training triplets. For each positive <user, outfit> pair, IQON-550 also provides 10 negative outfits, randomly sampled from other users' positive outfits. In this work, we randomly selected one from the 10 negative samples, and constituted a triplet <user, positive outfit, negative outfit>. Ultimately, we obtained 23,650 training and 8,250 validation triplets. As for testing, we adopted the original testing set provided by IQON-550, where for each user, there are 20 positive outfits with 200 negative ones for ranking. Each outfit in IQON-550 involves 3~8 items from different categories.

   Baselines. To validate the effectiveness of our proposed BiHGH scheme, we chose the following baselines for comparison.

- **BPR-MF** [23]. This is a matrix factorization model equipped with BPR loss, which projects the users and outfits into a latent space, and uses BPR loss to regulate the user's relative preference over the positive and negative outfits.
- **VBPR** [11]. It is a personalized fashion recommendation model based upon the matrix factorization framework, where the visual features of products are incorporated.
- **VBPR-T** [11]: This is extended from VBPR, which further incorporates the textual modality of products to enhance the item representation.

- **LPAE** [19]. This is a learnable personalized anchor embedding approach towards POR, which encodes the outfit with the self-attention mechanism, and models the user's preference with multiple anchors.
- **HFGN** [15]. This is a hierarchical fashion graph network devised for POR, which models the relationships among users, items, and outfits within a hierarchical graph.
- **FHN** [20]. This baseline aims to learn the binary hash code for each user and outfit towards efficient POR. In FHN, both visual and textual modalities of items are considered, and both BPR and VTE losses are jointly used.
- **FHN-T** [20]. Similar to VBPR-T, this baseline is extended from FHN, where the textual modality is incorporated.

It is worth noting that among these baselines, only FHN and FHN-T focus on learning hash codes of fashion entities, like ours.

Evaluation Tasks and Metrics. Towards comprehensive evaluation, we investigated two evaluation configurations: non-binary and binary. Specifically, in the former case, for hashing based methods (i.e., FHN, FHN-T, and ours), we used the continuous hash code of each user/outfit entity for evaluation. Regarding the non-hashing based methods, we judged them based upon their learned continuous entity representations. In the latter case, we directly imposed the $sign$ operation to convert the continuous representations to binary hash codes for hashing based methods in the testing phase. As for the other non-hashing based baselines, we re-trained them by deploying the same $tanh$ activation function over their fashion entity representations, and then tested them with the binary hash codes. For both configurations, we adopted two testing scenarios: binary preference prediction and personalized outfit recommendation. For the first testing scenario, we adopted the Area Under the ROC curve (AUC) [35] as the evaluation metric. Intuitively, AUC measures the probability that for a randomly selected pair of outfits for a give user, one positive and one negative, the predicted user preference score towards the positive outfit is higher than that towards the negative one. As for the personalized outfit recommendation, we adopted commonly used Mean Reciprocal Rank (MRR@K) [13] and Normalized Discounted Cumulative Gain (NDCG@K) [31, 38] as evaluation metrics. Specifically, MRR@K provides the insights into the ability of the model to return the positive outfit at top of the ranking list with K outfit candidates, and NDCG@K reflects the ranking positions of the positive outfits within the top K ranking list. In our work, we set $K = 10$.

Implementation Details. We utilized ResNet34 to derive the visual embedding of each item, whose final dimension is 512. Regarding the textual/attribute feature extraction, we resorted to BERT[2] for Japanese, since the dataset is in Japanese. The text/attribute feature outputted by BERT is a 768-dimensional vector. The GAT we adopted has 8 heads and 4 layers (i.e., $M = 8$ and $L = 4$). The dimension of the hash code is set to 512. As to the optimization, we adopted the adaptive moment estimation method. The learning rate is warmed up to the peak value, which is set to 5e-5, in the first 6% steps, and then linearly

---

[2] https://huggingface.co/cl-tohoku/bert-base-japanese-char/tree/main.

decayed to 0. We used the grid search strategy to derive the optimal hyperparameters. Ultimately, the batch size is set to 4. The margin hyperparameter c in contrastive loss is set to 0.1 and the weight hyperparameters α_1, α_2, and γ are set to 1, 1, and 0.04, respectively. All the experiments are implemented by PyTorch over a server equipped with 4 A100-PCIE-40GB GPUs.

6.4.2 On Model Comparison

Table 6.1 shows the performance comparison among different approaches with respect to three metrics. From Table 6.1, we have the following observations. (1) Our proposed BiHGH consistently outperforms all the baselines over all metrics across different evaluating configurations, which indicates the superiority of our scheme. (2) As compared to the non-binary setting, the performance of all the methods drops slightly in the binary evaluation setting. This is reasonable as there should be inevitable information loss during the binarization. Nevertheless, our model decreases the least. This may be attributed to that the dual similarity preserving regularization, which is beneficial to prevent the information loss in the hash layer. (3) Our method consistently surpasses all the baselines remarkably over all metrics under two settings, which implies the necessity of considering the attributes of items.

6.4.3 On Ablation Study

We conducted ablation study with the following derivatives. (1) **w/o text and w/o image**: To study the impact of the textual description and image of each fashion item for POR, we removed the their embeddings, respectively. (2) **w/o attribute**: To investigate the effect

Table 6.1 Performance comparison between our BiHGH and other baselines on IQON-550. The best results are in boldface, and the second best are underlined

Method	Non-binary			Binary		
	AUC	MRR	NDCG	AUC	MRR	NDCG
BPR-MF	0.6587	0.4735	0.5953	0.6259	0.4203	0.5546
VBPR	0.7491	0.5305	0.6424	0.7242	0.4999	0.6253
VBPR-T	0.7625	0.5463	0.6547	0.7354	0.5136	0.6293
LPAE	0.7204	0.5180	0.6316	0.7052	0.4899	0.6101
HFGN	0.7423	0.5091	0.6264	0.7264	0.5117	0.6281
FHN	0.7360	0.4950	0.6157	0.7297	0.4889	0.6108
FHN-T	0.7494	0.5109	0.6281	0.7426	0.5027	0.6217
BiHGH	**0.7974**	**0.5778**	**0.6799**	**0.7933**	**0.5742**	**0.6771**

Table 6.2 Ablation study results. The best results are highlighted in bold. @10 for MRR and NDCG is omitted due to the limited space

Method	Non-binary			Binary		
	AUC	MRR	NDCG	AUC	MRR	NDCG
w/o text	0.7911	0.5653	0.6704	0.7841	0.5578	0.6646
w/o image	0.7041	0.4842	0.6059	0.6978	0.4799	0.6025
w/o attribute	0.7911	0.5646	0.6698	0.7875	0.5675	0.6719
w/o DualSim	0.7902	0.5599	0.6658	0.7795	0.5335	0.6467
w/o VTE	0.7933	0.5686	0.6729	0.7850	0.5608	0.6669
w/o BiSeq	0.7825	0.5550	0.6622	0.7750	0.5537	0.6611
BiHGH	**0.7974**	**0.5778**	**0.6799**	**0.7933**	**0.5742**	**0.6771**

of the semantic attributes, we removed the attribute embeddings. (3) **w/o DualSim**: To justify the effectiveness of dual similarity preserving loss during the hash code learning, we eliminated both $\mathcal{L}_{S_o}$ and $\mathcal{L}_{S_u}$ from Eq. 6.11. (4) **w/o VTE**: To verify the VTE objective function, we removed the loss $\mathcal{L}_{VTE}$ from Eq. 6.11. (5) **w/o BiSeq**: To explore the effect of the bi-directional sequential graph learning, we replaced it with the general GAT over the graph $\hat{G}$ with four kinds of entities.

Table 6.2 summarizes the ablation study results. It can be seen that our model consistently outperforms all the above derivatives, which demonstrates the effectiveness of each component in our proposed BiHGH. Specifically, we have the following four detailed observations. (1) Both w/o text and w/o image perform inferior to BiHGH, which indicates that it is essential to consider both modalities of fashion items to boost the item representation learning. (2) w/o DualSim delivers the worse performance than our model, reflecting the benefit of the dual similarity preserving regularizations to prevent the information loss during the hash learning. (3) w/o VTE loss performs worse, which reveals that visual-semantic consistency regularizer largely benefits the hash code learning. (4) The performance of w/o BiSeq drops significantly, as compared to BiHGH, demonstrating that bi-directional sequential graph convolution is more powerful in transferring knowledge over the heterogeneous graph with various kinds of entities.

6.4.4 On Sensitivity Analysis

We also analyzed the sensitivity of our model regarding three key parameters: the number of training steps, the length D of hash codes, and the scale parameter β in the $sign$ function approximation.

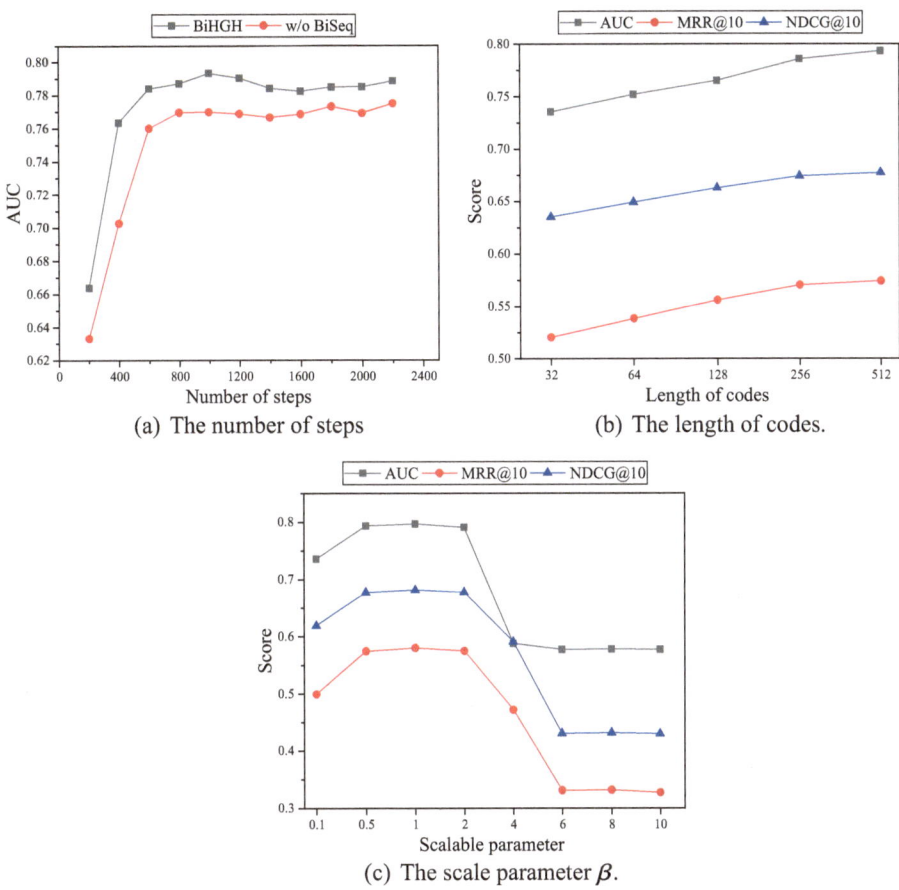

Fig. 6.4 Sensitivity analysis of our model in terms of the **a** number of steps, **b** length of codes, and **c** scale parameter β

On the Number of Training Steps. We comparatively plotted the detailed performance of our proposed BiHGH and the variant model w/o BiSeq on testing set at different training steps in Fig. 6.4a. We have two observations: (1) our model consistently outperforms w/o BiSeq at different training steps; and (2) our model converges faster as compared with w/o BiSeq. These observations validate the benefit of the bi-directional sequential graph convolution in promoting the entity representation learning and improving model convergence.

On the Length of Hash Codes. Figure 6.4b shows our model's performance with variable code lengths, i.e., {32, 64, 128, 256, 512} bits. As can be seen, with the increase of the code length, the performance of our model becomes better. This is reasonable, because longer hash codes convey more information.

On the Scale Parameter β. We varied this scale parameter among values of $\{0.1, 0.5, 1, 2, 4, 6, 8, 10\}$, and reported the corresponding performance in Fig. 6.4c. As can be seen, all the three metrics follow the same trend of first increasing until stable, and then sharply decreasing until stable again, with the increase of β. In particular, our model achieves the highest performance, when the scalar parameter β is set to 1. This implies that with the scale larger than 1, the elements of the continuous hash representations tend to be overly squeezed to 0 or 1, which leads to the severe information loss, and thus hurts the model performance.

6.4.5 On Case Study

To gain the intuitive understanding of our BiHGH, we conducted two case studies.

Firstly, we visualized the learned hash codes for users and outfits in the testing set by our model and its variant w/o DualSim with the tool of t-SNE [22]. The results are illustrated in Fig. 6.5, whereby different colors indicate different users and their corresponding positive outfits. From Fig. 6.5, we can see that our model is able to get the hash codes for the users and their preferred outfits closer than those obtained via w/o DualSim.

Secondly, we showed the top-4 ranking results of our method, and its derivatives w/o DualSim and w/o attribute for a given user in Fig. 6.6, where the users' historical preferred outfits are also listed to facilitate the experimental result analysis. As can be seen, our model outperforms its derivatives. Meanwhile, we noticed that as compared with w/o DualSim, our model is able to return more positive outfits that contain the same pair of shoes, highlighted in red boxes, which is preferred historically by the user. This confirms the effectiveness of our dual similarity preserving regularization. Furthermore, we observed that as compared with the w/o attribute, our model additionally returns a positive outfit at the fourth place. One possible explanation is that our model is able to uncover the user's preference over

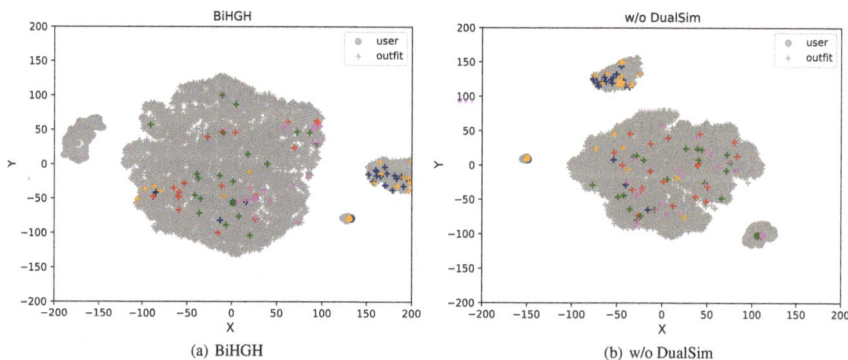

Fig. 6.5 Visualization of the learned hash codes of our BiHGH and its variant w/o DualSim

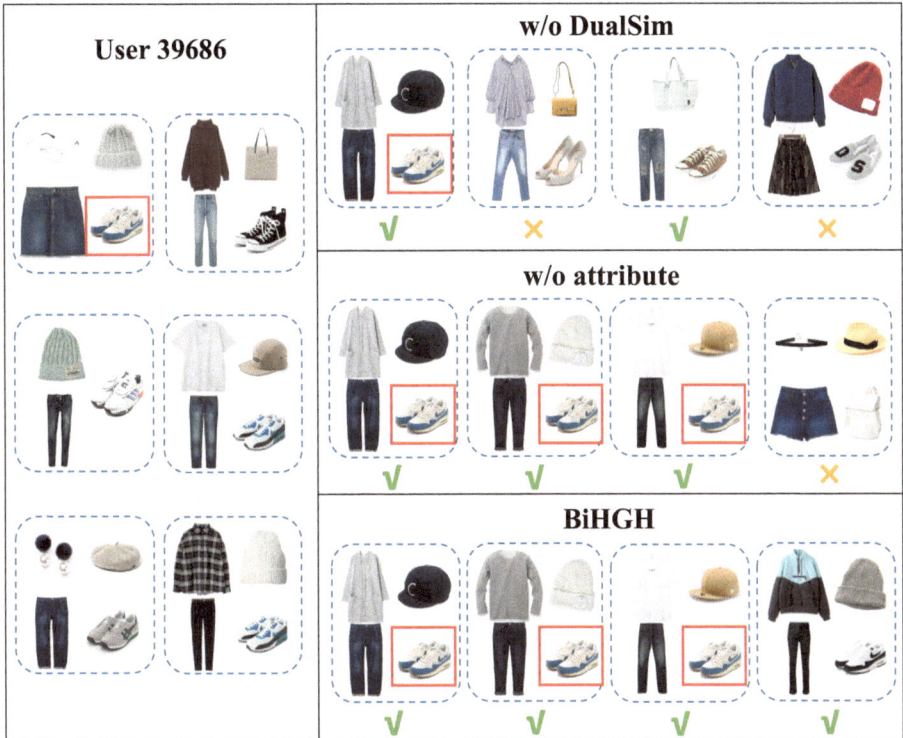

Fig. 6.6 Illustration of the POR ranking results obtained by our BiHGH, w/o DualSim, and w/o attribute

bottom items with the attribute "jeans", while the w/o attribute derivative fails. This also validates the benefit of incorporating the attribute information into the context of POR.

6.5   Conclusion and Future Work

In this work, we present an efficient personalized outfit recommendation scheme with Bi-directional heterogeneous graph hash learning, named BiHGH. Different from previous studies, we comprehensively unify the four types of entities (i.e., users, outfits, items and attributes) and their relations via a heterogeneous four-partite graph. Moreover, we creatively devise a bi-directional graph convolution algorithm to sequentially transfer knowledge via repeating upwards and downwards convolution to enhance the representation of users and outfits. We ultimately design the BPR loss for the user preference learning and dual similarity preserving regularization to prevent the information loss during the hash learning, respectively. Extensive experiments on the benchmark dataset demonstrate the superior performance and efficiency of BiHGH over existing methods. The ablation study also verifies the advantages of our bi-directional graph convolution over conventional ones, as well as the benefit of incorporating the dual similarity preserving regularization. One limitation of our work is that our current model cannot well deal with the cold-start problem in the outfit recommendation. In the future, we plan to address this issue.

References

1. Cao, Z., Long, M., Wang, J., Yu, P.S.: Hashnet: Deep learning to hash by continuation. In: Proceedings of the IEEE/CVF International Conference on Computer Vision, pp. 5609–5618. IEEE (2017)
2. Chen, W., Huang, P., Xu, J., Guo, X., Guo, C., Sun, F., Li, C., Pfadler, A., Zhao, H., Zhao, B.: Pog: personalized outfit generation for fashion recommendation at alibaba ifashion. In: Proceedings of the International ACM SIGKDD Conference on Knowledge Discovery and Data Mining, pp. 2662–2670 (2019)
3. Cui, H., Zhu, L., Li, J., Yang, Y., Nie, L.: Scalable deep hashing for large-scale social image retrieval. IEEE Transactions on Image Processing **29**, 1271–1284 (2020)
4. Devlin, J., Chang, M., Lee, K., Toutanova, K.: BERT: pre-training of deep bidirectional transformers for language understanding. In: Proceedings of the Annual Conference of the North American Chapter of the Association for Computational Linguistics, pp. 4171–4186. ACL (2019)
5. Faghri, F., Fleet, D.J., Kiros, J.R., Fidler, S.: VSE++: improving visual-semantic embeddings with hard negatives. In: Proceedings of the British Machine Vision Conference, pp. 12–12. BMVA Press (2018)
6. Gori, M., Monfardini, G., Scarselli, F.: A new model for learning in graph domains. In: Proceedings of the IEEE International Joint Conference on Neural Networks, pp. 729–734. IEEE (2005)
7. Guan, W., Wen, H., Song, X., Yeh, C.H., Chang, X., Nie, L.: Multimodal compatibility modeling via exploring the consistent and complementary correlations. In: Proceedings of the ACM International Conference on Multimedia, pp. 2299–2307 (2021)
8. Gune, O., Banerjee, B., Chaudhuri, S., Cuzzolin, F.: Generalized zero-shot learning using generated proxy unseen samples and entropy separation. In: Proceedings of the ACM International Conference on Multimedia, pp. 4262–4270. ACM (2020)
9. Han, X., Wu, Z., Jiang, Y.G., Davis, L.S.: Learning fashion compatibility with bidirectional LSTMs. In: Proceedings of the ACM International Conference on Multimedia, pp. 1078–1086 (2017)
10. He, K., Zhang, X., Ren, S., Sun, J.: Deep residual learning for image recognition. In: Proceedings of the IEEE/CVF Conference on Computer Vision and Pattern Recognition, pp. 770–778 (2016)
11. He, R., McAuley, J.: VBPR: visual bayesian personalized ranking from implicit feedback. In: Proceedings of the AAAI Conference on Artificial Intelligence, pp. 144–150. AAAI Press (2016)
12. Hou, M., Wu, L., Chen, E., Li, Z., Zheng, V.W., Liu, Q.: Explainable fashion recommendation: A semantic attribute region guided approach. In: Proceedings of the International Joint Conference on Artificial Intelligence, pp. 4681–4688 (2019)
13. Jiang, L., Yu, S., Meng, D., Yang, Y., Mitamura, T., Hauptmann, A.G.: Fast and accurate content-based semantic search in 100m internet videos. In: Proceedings of the ACM International Conference on Multimedia, pp. 49–58. ACM (2015)
14. Koren, Y., Bell, R.M., Volinsky, C.: Matrix factorization techniques for recommender systems. Computer **42**(8), 30–37 (2009)

15. Li, X., Wang, X., He, X., Chen, L., Xiao, J., Chua, T.S.: Hierarchical fashion graph network for personalized outfit recommendation. In: Proceedings of the International ACM SIGIR Conference on Research and Development in Information Retrieval, pp. 159–168 (2020)

16. Lin, Y., Moosaei, M., Yang, H.: Outfitnet: Fashion outfit recommendation with attention-based multiple instance learning. In: Proceedings of the ACM International Conference on World Wide Web Conference, pp. 77–87. ACM (2020)

17. Liu, H., Wang, R., Shan, S., Chen, X.: Deep supervised hashing for fast image retrieval. In: Proceedings of the IEEE/CVF Conference on Computer Vision and Pattern Recognition, pp. 2064–2072. IEEE (2016)

18. Lu, X., Zhu, L., Cheng, Z., Nie, L., Zhang, H.: Online multi-modal hashing with dynamic query-adaption. In: Proceedings of the International ACM SIGIR Conference on Research and Development in Information Retrieval, pp. 715–724. ACM (2019)

19. Lu, Z., Hu, Y., Chen, Y., Zeng, B.: Personalized outfit recommendation with learnable anchors. In: Proceedings of the IEEE/CVF Conference on Computer Vision and Pattern Recognition, pp. 12722–12731 (2021)

20. Lu, Z., Hu, Y., Jiang, Y., Chen, Y., Zeng, B.: Learning binary code for personalized fashion recommendation. In: Proceedings of the IEEE/CVF Conference on Computer Vision and Pattern Recognition, pp. 10562–10570. IEEE (2019)

21. McAuley, J., Targett, C., Shi, Q., Van Den Hengel, A.: Image-based recommendations on styles and substitutes. In: Proceedings of the International ACM SIGIR Conference on Research and Development in Information Retrieval, pp. 43–52 (2015)

22. Rauber, P.E., Falcão, A.X., Telea, A.C.: Visualizing time-dependent data using dynamic t-sne. In: Proceedings of the Eurographics Conference on Visualization, pp. 73–77. Eurographics Association (2016)

23. Rendle, S., Freudenthaler, C., Gantner, Z., Schmidt-Thieme, L.: BPR: bayesian personalized ranking from implicit feedback. In: Proceedings of the Conference on Uncertainty in Artificial Intelligence, pp. 452–461. AUAI Press (2009)

24. Sánchez, J., Perronnin, F.: High-dimensional signature compression for large-scale image classification. In: Proceedings of the IEEE/CVF Conference on Computer Vision and Pattern Recognition, pp. 1665–1672. IEEE (2011)

25. Song, X., Han, X., Li, Y., Chen, J., Xu, X.S., Nie, L.: Gp-bpr: Personalized compatibility modeling for clothing matching. In: Proceedings of the ACM International Conference on Multimedia, pp. 320–328 (2019)

26. Sun, C., Song, X., Feng, F., Zhao, W.X., Zhang, H., Nie, L.: Supervised hierarchical cross-modal hashing. In: Proceedings of the International ACM SIGIR Conference on Research and Development in Information Retrieval, pp. 725–734. ACM (2019)

27. Vedaldi, A., Zisserman, A.: Sparse kernel approximations for efficient classification and detection. In: Proceedings of the IEEE/CVF Conference on Computer Vision and Pattern Recognition, pp. 2320–2327 (2012)

28. Velickovic, P., Cucurull, G., Casanova, A., Romero, A., Liò, P., Bengio, Y.: Graph attention networks. In: Proceedings of the International Conference on Learning Representations, pp. 1–12. OpenReview.net (2018)

29. Wang, J., Zhang, T., Song, J., Sebe, N., Shen, H.T.: A survey on learning to hash. IEEE Transactions on Pattern Analysis and Machine Intelligence **40**(4), 769–790 (2018)

30. Wang, X., He, X., Cao, Y., Liu, M., Chua, T.: KGAT: knowledge graph attention network for recommendation. In: Proceedings of the International ACM SIGKDD Conference on Knowledge Discovery and Data Mining, pp. 950–958. ACM (2019)

31. Wang, X., He, X., Wang, M., Feng, F., Chua, T.S.: Neural graph collaborative filtering. In: Proceedings of the International ACM SIGIR Conference on Research and Development in Information Retrieval, pp. 165–174. ACM (2019)

32. Yang, X., Xie, D., Wang, X., Yuan, J., Ding, W., Yan, P.: Learning tuple compatibility for conditional outfit recommendation. In: Proceedings of the ACM International Conference on Multimedia, pp. 2636–2644 (2020)

33. Yu, W., Zhang, H., He, X., Chen, X., Xiong, L., Qin, Z.: Aesthetic-based clothing recommendation. In: Proceedings of the ACM International Conference on World Wide Web Conference, pp. 649–658. ACM (2018)

34. Zhang, H., Liu, M., Gao, Z., Lei, X., Wang, Y., Nie, L.: Multimodal dialog system: Relational graph-based context-aware question understanding. In: Proceedings of the ACM International Conference on Multimedia, pp. 695–703. ACM (2021)

35. Zhang, H., Zha, Z.J., Yang, Y., Yan, S., Gao, Y., Chua, T.S.: Attribute-augmented semantic hierarchy: towards bridging semantic gap and intention gap in image retrieval. In: Proceedings of the ACM International Conference on Multimedia, pp. 33–42 (2013)

36. Zheng, N., Song, X., Niu, Q., Dong, X., Zhan, Y., Nie, L.: Collocation and try-on network: Whether an outfit is compatible. In: Proceedings of the ACM International Conference on Multimedia, pp. 309–317. ACM (2021)

37. Zhou, K., Zha, H.: Learning binary codes for collaborative filtering. In: Proceedings of the International ACM SIGKDD Conference on Knowledge Discovery and Data Mining, pp. 498–506. ACM (2012)

38. Zhou, Z., Di, X., Zhou, W., Zhang, L.: Fashion sensitive clothing recommendation using hierarchical collocation model. In: Proceedings of the ACM International Conference on Multimedia, pp. 1119–1127. ACM (2018)

39. Zhu, H., Long, M., Wang, J., Cao, Y.: Deep hashing network for efficient similarity retrieval. In: Proceedings of the AAAI Conference on Artificial Intelligence, pp. 2415–2421. AAAI Press (2016)

Diverse Collocated Clothing Synthesis for Outfit Recommendation

7.1   Introduction

In this chapter, we explore the synthesis of collocated clothing for outfit recommendation. Recommending visually-collocated outfits holds substantial value for the fashion industry and significantly enhances the business prospects of e-commerce platforms. For example, in recent years, there have been several proliferating platforms such as iFashion, ZOZO, and Xiaohongshu, which have launched their fashion matching applications, aiming to improve the turnover of clothing. As a result, this has also attracted considerable attention from the academic community. Researchers have carried out a series of studies on learning fashion compatibility, usually leveraging algorithms to determine whether a composed outfit is collocated or not [7, 11, 24, 27, 29, 40, 41]. The focus of these works, however, mainly lies in finding how to recommend appropriate fashion items that are compatible with the ones bought by customers, conditioned on the existence of the recommended fashion items in a database. For learning fashion compatibility, the flip side of the coin is to help fashion designers create new collocated apparel. Since generative adversarial networks (GANs) [9, 36, 46] have gradually become prominent in image synthesis, they have the potential to provide design inspirations for fashion designers through generating large pools of new garments which are compatible with existing ones [25, 26, 45, 48, 50]. Specifically, for generating collocated clothing generation, Liu et al. [25] first used a GAN framework, named Attribute-GAN, to assist designers in creating new apparel from a fashion compatibility learning perspective. In their model, a compatible upper clothing item can be synthesized based on a given lower clothing item, or vice versa. Liu et al. [26] then extended Attribute-GAN with a multi-discriminator architecture to improve the synthesis quality. The above studies, however, overlooked the personalized needs of users for clothing collocation. To address this limitation, Yu et al. [45] developed a collocated clothing generation framework for a personalized design. Their method works on the collocated clothing generation tasks between upper and lower clothing domains with user preference features. Although these

W. Guan et al., *Advanced Multimodal Compatibility Modeling and Recommendation*, Synthesis Lectures on Information Concepts, Retrieval, and Services, https://doi.org/10.1007/978-3-031-81048-0_7

works focused on only learning the mapping between upper and lower clothing domains, they shed light on the feasibility of collocated clothing synthesis toward intelligent design and even paved the way to outfit generation using GANs [48, 50]. Nevertheless, three main drawbacks of these studies remain to be solved: (i) they can only synthesize a single image of compatible clothing at a time based on a given clothing item; (ii) these methods mainly rely on general image-to-image (I2I) translation frameworks, which may be ineffective in rendering images across clothing categories due to the large spatial non-alignment of the semantics between the source and target domains; and (iii) they usually require additional information to guide the image generation, e.g., the attributes of the clothing [25, 26], collected data of user preferences [45], or reference masks [48, 50]. For the first issue, in practice, it is more desirable to provide multiple, diverse synthesized results rather than only one result to fashion designers: this will allow them to select their favorite ones, which can be further elaborated and used in their target design items. For the second issue, the existing I2I translation methods cannot be simply adapted to perform the generation of compatible fashion items, as there is no apparent spatial alignment between them. In response to the third issue, a more flexible generation model without prior conditions should be developed, as users may have difficulties acquiring the additional information as model input in real-world applications. In particular, a model that has to synthesize multiple images at one time can be seen as a kind of multimodal I2I translation method [5, 15, 22, 23, 28] residing in fashion domains. Generally, the aim of current multimodal I2I translation methods is to learn a distribution that explores multiple possible images in the target domain based on a given image from the source domain as input. Recently, a series of studies [5, 15, 22, 23, 28] have been explored for multimodal I2I translation methods. Unfortunately, when these methods are directly applied to visually-collocated and diverse clothing synthesis, they often encounter their Waterloos. The rationale behind this may be ascribed to the following facts. Unlike the problem setting of these general methods, two crucial issues are faced in our task: (i) the translation of 'upper $\rightleftharpoons$ lower' clothing has no explicit spatial alignment, in contrast to other general I2I translations, e.g., 'summer $\rightleftharpoons$ winter' translation there actually is a spatial alignment in pixel level; and (ii) since general multimodal I2I translation methods usually have no ground truths in the target domain, they can only learn the mapping from a source domain to a target domain in an unsupervised manner. For the first issue, these multimodal I2I methods find it hard to resolve the problem of semantic non-alignment due to the fact that they usually adopt an encoder–decoder (or U-Net) architecture, making it impossible for them to be directly applied to our task. For the second issue, our task constitutes many possible ground truths naturally, i.e., matching pairs of upper and lower clothing, which should be fully used to guide the clothing synthesis process. To properly address the above issues in our task, we propose a new framework, named *BC-GAN*, to synthesize multiple compatible clothing simultaneously. Our model is built upon a pre-trained model, which helps us encode images into embeddings lying in the pre-trained latent space. The embeddings are fed into a pre-trained model to produce the targeted images. Our model is able to address the issues of spatial semantic non-alignment by taking advantage

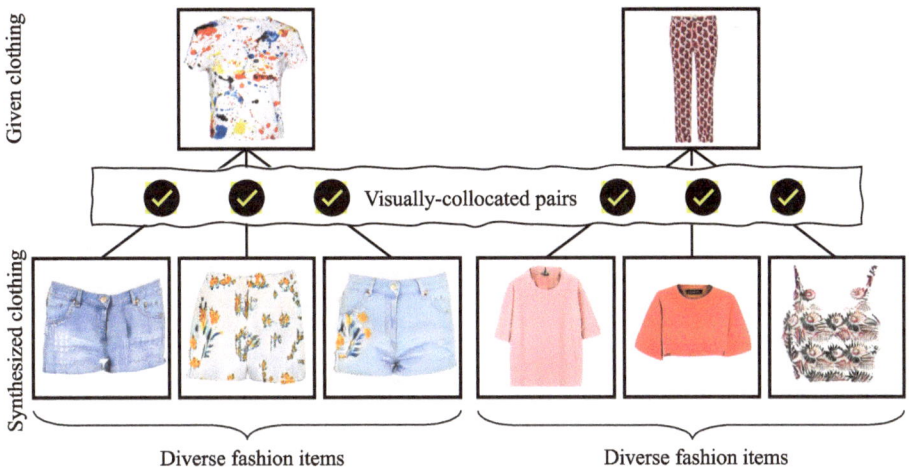

Fig. 7.1 Multiple visually-collocated fashion items synthesized by our BC-GAN

of the formulated embeddings and the generation ability of the pre-trained model. We also develop a new compatibility discriminator with a supervision ability to further improve the fashion compatibility of the synthetic results in comparison to that of previous studies [25, 26, 45]. A large-scale dataset containing 31,631 outfits was constructed to train and evaluate the proposed model. Extensive experiments were conducted to validate its effectiveness with respect to various evaluation metrics, in comparison to current state-of-the-art methods. Figure 7.1 illustrates certain samples synthesized by our BC-GAN. For each given fashion item, BC-GAN can produce a set of diversified items. In particular, visually-collocated pairs of fashion items have been composed with a given item in both 'upper → lower' and 'lower → upper' directions.

To sum up, the main contributions of this research are:

1. To the best of our knowledge, we are the first to develop a generation framework, named BC-GAN, to synthesize multiple collocated fashion items at one time. By leveraging the power of a pre-trained model, BC-GAN can accurately learn the mapping between the upper and lower clothing domains, overcoming the spatial non-alignment of the semantics during translation.

2. We develop a contrastive learning-based compatibility discriminator to further improve the compatibility between the given and the synthesized fashion items. Specifically, their compatibility relations are modeled from a contrastive learning perspective, in order to further enrich the supervision information for the developed discriminator.

3. We construct a large-scale fashion dataset *DiverseOutfits*, which consists of a large number of compatible and diverse outfits. Extensive experiments validate that our proposed method can synthesize multiple compatible clothing based on a given item of clothing. Qualitative and quantitative results confirm that BC-GAN generates better results

than the state-of-the-art methods in terms of diversity, visual authenticity, and fashion compatibility.

The remainder of this chapter is structured as follows. Section 7.2 reviews the related work, followed by the introduction of the proposed BC-GAN in Sect. 7.3. Section 7.4 details the construction of the dataset and presents the experimental results. Finally, some conclusions are drawn from our research in Sect. 7.5.

7.2   Related Work

Our work is most related to three streams of research, i.e., fashion compatibility learning, multimodal I2I translation, and GAN inversion. In this section, we first briefly review the research related to these three realms. Then we highlight the features of this work in comparison to the existing ones.

Fashion Compatibility Learning. Fashion compatibility learning aims to address the collocation issues of a composed outfit. In general, current studies can be categorized into two sectors: discriminative models and generation models. For the first sector, many studies [7, 11, 24, 29, 40, 41] focused on determining whether an outfit is compatible or not by exploring deep learning models. McAuley et al. [29] first used a pre-trained convolutional neural network (CNN) to extract the features of fashion items, and then calculated their in-between distance to evaluate their compatibility. At the same time, Veit et al. [41] adopted Siamese CNNs to formulate better style representations for fashion items. Later, in the work of [40], a type-aware network, was proposed to more accurately model fashion compatibility. These methods which are dependent on metric learning [6], attempted to encode all fashion items into feature embeddings to facilitate the calculation of the distance between pairs of items in an outfit. On the other hand, Han et al. [11] argued that fashion compatibility can be modeled in a sequence model from the perspective of human observation. They adopted a bidirectional long short-term memory (Bi-LSTM) network [34] to improve the measuring of fashion compatibility. In recent years, graph neural networks (GNNs) [7, 24] as emerging topics in computer vision, have been employed to characterize the relationships among fashion items. Apart from the aforementioned discriminative models, many investigations [25, 26, 45, 48, 50] have paid considerable attention to synthesizing visually-collocated clothing by generation models, such as GAN [9]. These methods regard the synthesis of collocated clothing as an I2I translation task. Specifically, they attempted to synthesize the images of complementary fashion items based on those of the given items. For example, Liu et al. [25] first developed an Attribute-GAN framework to synthesize compatible fashion items based on a given item using attribute annotations with 'upper → lower' and 'lower → upper' directions. Subsequently, they further extended Attribute-GAN with a multi-discriminator architecture to improve its collocated clothing synthesis [26]. In contrast, Yu et al. [45] developed a model that used the data associated with user preferences regarding outfits to facilitate personalized design for customers. In particular, these studies only worked on the translation between upper and lower clothing images to perform fashion collocation. Recently, Zhou et al. [48, 50] developed frameworks for synthesizing a set of complementary fashion items to compose an outfit with given items. However, all of these studies constitute unimodal frameworks that can only produce one complementary clothing image at a time.

Multimodal I2I Translation. Multimodal I2I translation has the task of learning a conditional distribution that explores multiple possible images in a target domain, given an input image in a source domain. In order to produce diverse results, multimodal methods usually take randomly sampled noises as additional inputs. In particular, Huang et al. [15] proposed a multimodal I2I translation framework, called MUNIT, to disentangle images into style and content codes in a high-level feature space. During the same period, Lee et al. [22] proposed a disentangled representation framework for I2I translation (DRIT) to synthesize diverse images based on the same input. In line with this study, an improved version, named DRIT++ [23], extended the original framework to multiple domains for synthesizing better results. Furthermore, Choi et al. [5] proposed a framework with a star-shape translation network for synthesizing diverse results, especially for human faces or animal faces. More recently, a novel model [28], named SAVI2I, was proposed to achieve a continuous translation between source and target domains by using signed vectors. All of these methods were trained in an unsupervised manner due to the lack of ground truths in the investigated task.

GAN Inversion. GAN inversion [31, 39, 42, 51] is a technique that inverts an image back into the latent space of a pre-trained GAN model as a latent code so that the visual attribute of a given image can be manipulated by editing the inverted latent code. Generally, inversion methods include optimizing the latent codes to minimize the loss for the given images [38], training an encoder to map the given images into the latent space [31, 39, 51], or use a hybrid method combining both strategies [2]. The inverted latent codes can be edited by finding linear directions that correspond to changes in a given binary attribute, such as 'young ↔ old', or 'male ↔ female' [31, 39, 51]. These mentioned methods manipulate the attribute of the given images by editing the inverted latent code with a pre-trained support vector machine (SVM) [35] and can be only performed in the same image domain. The fundamental principle behind GAN inversion is the disentanglement of the space of a pre-trained StyleGAN in the dimension of visual image attributes. Consequently, the inverted latent codes can be manipulated via a pre-trained SVM. It is worth noting that these inverted latent codes can only be edited and subsequently fed into the pre-trained generator for unimodal I2I translation to accomplish visual attribute editing.

Positioning of Our Work. Among the studies discussed above, the closest works to ours are the models proposed by [25, 26], and [45]. Our task is expected to take an image of given clothing as input and produce diverse and compatible images of clothing, e.g., 'upper → lower'. It falls in the category of multimodal I2I translation methods. Most multimodal I2I translation methods [5, 15, 22, 23, 28] usually generate diverse images at one time in an unsupervised learning setting, and the semantics of inputs and outputs of these models have an explicit spatial alignment. For example, the 'summer ⇌ winter' translation [15, 22,

23, 28] relies on pixel-level spatial alignment, while the human face translation [5, 22, 23, 28] is implemented by hidden landmark alignment. Unfortunately, the translation task in this research, 'upper $\rightleftharpoons$ lower' clothing, cannot be based on this due to the lack of explicit spatial semantic alignment between upper and lower clothing items. To alleviate the spatial nonalignment between the input and output in the task of diverse and collocated clothing synthesis. We adopt GAN inversion technique to encode images into vectors to ignore the effect of spatial information. It should be noted here that our framework is different from the GAN inversion from three aspects: (i) the composed components of our BC-GAN is different from those of GAN inversion, the additional pre-trained SVM [35] to perform visual attribute manipulation is not needed in our framework; (ii) the motivation of our BC-GAN is different from that of GAN inversion. Our aim is to encode images into vectors to ignore the effect of spatial information but the insight of GAN inversion is the $\mathcal{W}$ space is disentangled about visual attributes; and (iii) the GAN inversion only supports unimodal I2I translation in visual attributes but our BC-GAN supports multimodal I2I translation in collocated clothing synthesis. In addition, current multimodal I2I translation methods can only use unsupervised strategies to train their models, as there are no ground truths in a training dataset. On the contrary, the availability of existing matching pairs of clothing items can be fully exploited by transforming them as a type of weakly supervised information in comparison with those unsupervised methods.

7.3   BC-GAN

In this section, we first formulate our research problem formally. Afterward, to elaborate on the key elements in the design of our framework, we discuss related network architectures, including StyleGAN [18] and its variants [31, 39, 51]. Then, we introduce the overall framework of BC-GAN. The training objectives and the implementation details for training our model are then illustrated.

7.3.1   Problem Formulation

In general, the focus of previous collocated clothing synthesis methods [25, 26, 45] is leveraging additional condition information to synthesize unimodal fashion items which are expected to be compatible with given items. To achieve this, such methods usually employ an encoder–decoder (or U-Net) architecture to generate visually-collocated clothing. In contrast, recent pre-trained generative models enable us to effectively overcome the issues of semantic non-alignment and synthesize multimodal compatible fashion items. By taking advantage of these pre-trained models, our task aims at synthesizing diverse fashion items, which are compatible with a given fashion item, without using any additional information as input. As a specific multimodal I2I translation task in the fashion domain, at the current stage, we work on two translation directions, i.e., 'upper → lower' and 'lower → upper'. This means that given an upper (or lower) clothing item, our framework can learn a mapping capable of synthesizing multiple different lower (or upper) clothing items that are compatible with the given one. To ensure our framework can synthesize diverse results, random noise is needed to feed into our model, as widely performed in previous studies [15, 22, 23, 28]. Formally, a mapping function in I2I translation can be represented as $F : \mathcal{X} \times \mathcal{Z} \to \mathcal{Y}$, where $\mathcal{X}$, $\mathcal{Z}$ and $\mathcal{Y}$ denote a source image domain, a given noise distribution and a target image domain, respectively; and $\times$ is the Cartesian product. Take the 'upper → lower' clothing translation as an example in this research, $\mathcal{X}$ and $\mathcal{Y}$ are the upper and lower clothing domains, respectively; lying in $\mathbb{R}^{H \times W \times 3}$, where H and W denote the height and width of the images, respectively. $\mathcal{Z}$ lies in $\mathbb{R}^L$, where L is the dimension of each latent code in the latent space. The translation 'lower → upper' follows a similar definition. The optimization objective of our task constitutes three perspectives on the generation results, namely, diversity, visual authenticity, and fashion compatibility. Formally, given a set of n latent codes $\mathbf{z}_1, \dots, \mathbf{z}_n \in \mathcal{Z}$ and a given source clothing item $\mathbf{x} \in \mathcal{X}$, the mapping function becomes $F : \{\mathbf{x}, \mathbf{z}_i\} \mapsto \mathbf{y}_i$, where $\mathbf{z}_i$ is the i-th sampled noise and $\mathbf{y}_i$ is the i-th synthesized result. In essence, we argue that an ideal framework for diverse and compatible clothing generation should make $\mathbf{y}_i$ fit the distribution of $\mathcal{Y}$, make $\mathbf{x}$ and $\mathbf{y}_i$ compose a visually-collocated outfit, and make $\{\mathbf{y}_i\}_{i=1}^n$ possess a large variance in terms of a human visual perspective.

7.3.2 Pre-trained Models for BC-GAN

As mentioned, previous methods on multimodal image translation [5, 15, 22, 23, 28] have difficulties addressing semantic misalignment issues due to the use of encoder-decoder architectures. To address this, we propose a new generation framework, namely, BC-GAN, which is built upon a pre-trained StyleGAN [18]. For clarity, we first briefly introduce StyleGAN and its variants [31, 39, 49, 51]. We will then propose the generation module of our framework and further highlight the features of our work in comparison to existing studies.

The generator architecture of StyleGAN adopts adaptive style manipulation and progressive growth to synthesize photo-realistic images. As shown in Fig. 7.2a, StyleGAN designed a generator for unconditional image generation with two key components: a mapping network f and a synthesis network g. The mapping network f constitutes a multi-layer perceptron (MLP) with eight layers. Here, f is responsible for mapping a latent code $\mathbf{z} \in \mathbb{R}^{512}$, sampled from a Gaussian distribution to a style embedding $\mathbf{w}$ lying in the space $\mathcal{W}$, where $\mathcal{W}$ is a disentangled latent space. Then, the style embedding $\mathbf{w}$ is fed into the synthesis network g to obtain an image I. By now, the rapid evolution of the StyleGAN series [17–19] has spawned many related studies [31, 39, 51], which have tried to understand and control the learned latent spaces, in order to achieve accurate attribute manipulation. As shown in

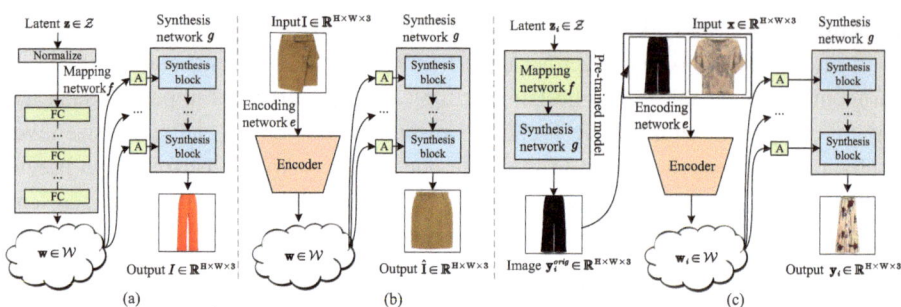

Fig. 7.2 Comparable frameworks of the StyleGAN-based models: **a** StyleGAN [18], **b** GAN inversion for edit framework [51], and **c** BC-GAN (ours). Here, the pre-trained model contains mapping network f and synthesis network g

Fig. 7.2b, Zhu et al. [51] proposed an additional encoding network e to encode images beyond the pre-trained model. Their training strategy is similar to that of variational auto-encoders (VAEs) [21] in a self-supervised manner. During the training phase, their encoders were trained with supervised loss functions, e.g., L1 loss [16], learned perceptual image patch similarity (LPIPS) [47] loss, and identification (ID) loss [8]. During the test phase, these methods aim to perform I2I translation by adopting the key idea of "encoding first, edit later" [31, 39, 51]. Images are first encoded into a style embedding $\mathbf{w}$ using the above-mentioned encoder e. Then an attribute edit can be performed by moving $\mathbf{w}$ along a certain direction in the $\mathcal{W}$ space. It is worth noting that some methods [31, 39] use an extended space, $\mathcal{W}+$, to obtain the edited style embedding $\hat{\mathbf{w}}$, which is then fed into the pre-trained synthesis network g to generate an edited image $\hat{I}$. However, they cannot be directly applied to our diverse and compatible clothing synthesis task. We attribute the underlying reason for this to the following three factors. First and foremost, the strategy for training such an encoder needs one-to-one mapping pairs, which is unachievable in our research problem, because an upper clothing item may be compatible with multiple lower clothing items according to different dressing scenarios and personal preferences. Besides, in the 'upper $\rightleftharpoons$ lower' task, the edit operation performed on the style embedding cannot be implemented in the clothing translation domain due to the necessity of training an edit network in a single domain. Last but not least, our task aims at realizing multimodal I2I translation in clothing domains, while those existing frameworks only support unimodal image generation.

To leverage the power of a pre-trained model, we propose a new paradigm for multimodal I2I translation, which is specialized for working on multimodal 'upper $\rightleftharpoons$ lower' translation tasks. In particular, the generator of our framework, G, consists of three key components, including a mapping network f, an encoding network e, and a synthesis network g, as illustrated in Fig. 7.2c. Specifically, for a batch of random latent codes $\{\mathbf{z}_i\}_{i=1}^{n}$ sampled from a normal Gaussian distribution $\mathcal{N}(0, \mathbf{I})$, we first feed them into a pre-trained model to obtain diverse images lying in the target domain, which can be formulated as $f \circ g(\mathbf{z}_i) \mapsto \tilde{\mathbf{y}}_i$, where $1 \leq i \leq n$. Then we concatenate this image $\tilde{\mathbf{y}}_i$ and a given image $\mathbf{x}$ from domain $\mathcal{X}$. Next, the encoder e is used to encode the above-concatenated images into a style embedding $\mathbf{w}$ lying in $\mathcal{W}$. The encoded embedding $\mathbf{w}$ is fed into the pre-trained synthesis network g, in order to synthesize an image $\mathbf{y}_i$ lying in the domain $\mathcal{Y}$. With this design, the generation framework can effectively support multimodal I2I translation and leverage the power of the pre-trained model, as we take advantage of the generation ability of the pre-trained model.

7.3.3   Proposed Framework

In line with the basic generation architecture of the proposed BC-GAN, in this subsection, we introduce the detailed architecture of our BC-GAN, which is elaborated in Fig. 7.3. The optimization process of BC-GAN will be detailed in Sects. 7.3.4 and 7.3.5. In particular, BC-GAN consists of three key modules described in the following.

Generator. The generator G shown in Fig. 7.3a is tasked with translating an input image $\mathbf{x}$ and a set of sampled latent codes $\{\mathbf{z}_i\}_{i=1}^{n}$ into a set of output images $\{\mathbf{y}_i\}_{i=1}^{n}$. It contains three key components, as described in Sect. 7.3.2, namely, a mapping network f, an encoding network e, and a synthesis network g. In particular, f and g were pre-trained on the target domain $\mathcal{Y}$ using the training set. The detailed encoding network of our e is constructed upon that of pSp [31]. Formally, our generator is tasked with synthesizing a set of images $\{\mathbf{y}_i\}_{i=1}^{n}$. For each input image $\mathbf{x}$ and a given random noise $\mathbf{z}_i$, a new image $\mathbf{y}_i$ can be synthesized by the generator. The detailed feed-forward of $G(\mathbf{x}, \mathbf{y}_i)$ is described in the following three steps. First, each latent code $\mathbf{z}_i$ is fed into f to obtain the original style

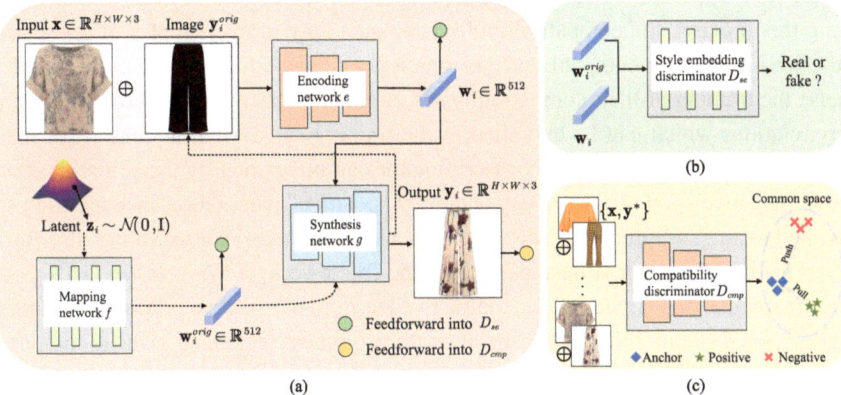

Fig. 7.3 Key components of our proposed BC-GAN: **a** The generator G which translates an input image $\mathbf{x}$ and a randomly sampled latent code $\mathbf{z}_i$ into an output image $\mathbf{y}_i$, **b** the style embedding discriminator D_{se} which distinguishes between real and fake style embeddings, and **c** the compatibility discriminator D_{cmp} which provides fashion compatibility supervision for synthesized clothing images in a contrastive learning perspective

embedding $\mathbf{w}_i^{orig}$ in the $\mathcal{W}$ space; $\mathbf{w}_i^{orig}$ is then fed into g for synthesizing an original image $\mathbf{y}_i^{orig}$ lying in domain $\mathcal{Y}$. Second, the input image $\mathbf{x}$, which is concatenated with $\mathbf{y}_i^{orig}$ in channel dimension, is fed into e to obtain the encoded style embedding $\mathbf{w}_i$. Finally, $\mathbf{w}_i$ is fed into g for synthesizing our targeted image $\mathbf{y}_i$. Formally, for each synthesized $\mathbf{y}_i$, it can be obtained by $g(e(\mathbf{x}\oplus(f \circ g(\mathbf{z}_i)))) \mapsto \mathbf{y}_i$, where $\oplus$ is a concatenation operation in channel dimension. It is worth noting that our generator also supports continuous translation by using linear interpolation between $\mathbf{w}_i^{orig}$ and $\mathbf{w}_i$.

Style Embedding Discriminator. To ensure the visual authenticity of the synthesized results of BC-GAN, a real/fake discriminator needs to be developed to supervise the generator. In fact, we have two choices for achieving this goal: (i) employing a network to supervise the generated images fitting the distribution of $\mathcal{Y}$, or (ii) employing a network to supervise the encoded style embeddings fitting the distribution of $\mathcal{W}$. In particular, the second alternative which has been adopted in this research allows a model to keep the disentanglement characteristic of the encoded style embedding. Moreover, it is more efficient during model training, as the use of embeddings only needs to calculate the real/fake loss in $\mathbb{R}^L$ rather than using the final output images with a computational requirement in $\mathbb{R}^{H \times W \times 3}$. Here, we present a style embedding discriminator D_{se} illustrated in Fig. 7.3b to ensure the visual authenticity, due to the fact that the latent code $\mathbf{w}_i$ in the $\mathcal{W}$ space can favor the implementation of continuous translation in the target domain (see Sect. 7.4.8). Our style embedding discriminator consists of an MLP with four layers. To keep the training process more stable, we also used a discriminator pool to preserve style embeddings. This discriminator is responsible for supervising the encoded style embeddings in the $\mathcal{W}$ space.

Contrastive Learning-Based Compatibility Discriminator (Figure 7.3c). For the design of the fashion compatibility discriminator, previous studies [25, 26, 45] adopted a scheme by concatenating upper and lower clothing as a tuple which is fed into a discriminator directly. During this process, the compatibility discriminator only receives tuples of real pairs and synthesized pairs. The compatibility information cannot be fully exploited in this design, because the training of their compatibility discriminators has not considered the incompatibility relations which can be exploited in the training set. Based on contrastive learning [3, 10, 33, 37, 43], a compatibility discriminator can be learned by using a representation to push "negative" pairs further away and draw "positive" pairs closer in a common space. Motivated by this, we propose a new contrastive learning-based compatibility discriminator D_{cmp}. This discriminator is designed under the framework of Wasserstein GAN [1]. The discriminator can be learned from a contrastive learning perspective. In previous studies [7, 11, 24, 29, 40, 41], the compatible pairs of clothing items were defined as those composed by fashion experts, whereas the incompatible pairs of clothing were defined as those composed with random combinations. In our implementation, all pairs for training the compatibility discriminator can be divided into three types, $\{\mathbf{x}, \mathbf{y}^{r,c}\}$, $\{\mathbf{x}, \mathbf{y}^{r,!c}\}$, and $\{\mathbf{x}, \mathbf{y}^{f,c}\}$, where $\mathbf{x}$ is the given clothing from $\mathcal{X}$, $\mathbf{y}^{r,c}$ is a sampled item of clothing from $\mathcal{Y}$ which is compatible with $\mathbf{x}$, $\mathbf{y}^{r,!c}$ is a sampled clothing item which is not compatible with $\mathbf{x}$, and $\mathbf{y}^{f,c}$ is an item of clothing synthesized by BC-GAN, which is expected to be compatible with $\mathbf{x}$. In particular, the former two sampled items of clothing in the target domain are all selected from the training set. The training of the compatibility discriminator should ensure that the discriminator can distinguish incompatible and compatible pairs as much as possible. In contrast, all pairs of items for training our generator can be divided into three types, $\{\mathbf{x}, \mathbf{y}^{r,c}\}$, $\{\mathbf{x}, \mathbf{y}^{f,c}\}$, and $\{\mathbf{x}, \mathbf{y}^{f,!c}\}$, where $\mathbf{y}^{f,!c}$ is an item synthesized by our framework, which is not expected to be compatible with $\mathbf{x}$. Each composed pair is fed into the compatibility discriminator to obtain an embedding in a common space. For training the compatibility discriminator, we expect that the embedding of $\{\mathbf{x}, \mathbf{y}^{r,c}\}$ in the space is far from those of $\{\mathbf{x}, \mathbf{y}^{r,!c}\}$ and $\{\mathbf{x}, \mathbf{y}^{f,c}\}$. For training the generator with the compatibility loss provided by the aforementioned discriminator, the embedding of $\{\mathbf{x}, \mathbf{y}^{f,c}\}$ in the space is expected to be close to that of $\{\mathbf{x}, \mathbf{y}^{r,c}\}$, and to be far from that of $\{\mathbf{x}, \mathbf{y}^{f,!c}\}$.

7.3.4 Training Objectives

Given a batch of random latent codes $\{\mathbf{z}_i\}_{i=1}^n$ sampled from $\mathcal{Z}$ and an image $\mathbf{x} \in \mathcal{X}$, we train our framework using the following objectives.

Real/Fake Objective. During training, we sample a batch of latent codes $\mathbf{z}_1, \ldots, \mathbf{z}_i, \ldots,$ $\mathbf{z}_n \in \mathcal{Z}$ randomly and are given an image from $\mathbf{x} \in \mathcal{X}$, and generate two style embeddings: $\mathbf{w}_i^{orig} = f(\mathbf{z}_i)$ and $\mathbf{w}_i = e(\mathbf{x} \oplus f \circ g(\mathbf{z}_i))$. The style embedding discriminator D_{se} takes these style embeddings $\mathbf{w}_i^{orig}$ and $\mathbf{w}_i$ as inputs and learns to distinguish the original style embeddings and the encoded style embeddings with the following loss:

$$
\begin{aligned}
\mathcal{L}_{dis} =& \mathbb{E}_{\mathbf{z}_i \sim \mathcal{Z}, \mathbf{x} \sim \mathcal{X}} [\log D_{se}(f(\mathbf{z}_i)) + \\
& \log(1 - D_{se}(e(\mathbf{x} \oplus f \circ g(\mathbf{z}_i))))],
\end{aligned}
\tag{7.1}
$$

where two terms on the right hand are responsible for $\mathbf{w}_i^{orig}$ and $\mathbf{w}_i$, respectively. The encoder e takes an image $\mathbf{x}$ and a random synthesized image $f \circ g(\mathbf{z}_i)$ as inputs and learns to generate a style embedding via an adversarial loss:

$$
\mathcal{L}_{adv} = \mathbb{E}_{\mathbf{z}_i \sim \mathcal{Z}, \mathbf{x} \sim \mathcal{X}} [\log(D_{se}(e(\mathbf{x} \oplus f \circ g(\mathbf{z}_i))))].
\tag{7.2}
$$

With these training losses, our generator can produce $\mathbf{w}_i$ lying in $\mathcal{W}$ space. This real/fake objective is designed to guide our generator to synthesize photo-realistic images.

Fashion Compatibility Objective. To guarantee the compatibility of a synthesized clothing item $\mathbf{y}_i$ with the given image $\mathbf{x}$, we employ a contrastive loss for training our compatibility discriminator and our generator. For training the compatibility discriminator D_{cmp}, we have the following training objective:

$$\mathcal{L}_{cmp\_dis} =$$
$$[\mathbb{E}_{\mathbf{z}_i \sim \mathcal{Z}, \mathbf{x} \sim \mathcal{X}}(D_{cmp}(\{\mathbf{x}, \mathbf{y}^{r,c}\}))$$
$$-\mathbb{E}_{\mathbf{z}_i \sim \mathcal{Z}, \mathbf{x} \sim \mathcal{X}}(D_{cmp}(\{\mathbf{x}, \mathbf{y}^{r,!c}\}))] \tag{7.3}$$
$$+[\mathbb{E}_{\mathbf{z}_i \sim \mathcal{Z}, \mathbf{x} \sim \mathcal{X}}(D_{cmp}(\{\mathbf{x}, \mathbf{y}^{r,c}\}))$$
$$-\mathbb{E}_{\mathbf{z}_i \sim \mathcal{Z}, \mathbf{x} \sim \mathcal{X}}(D_{cmp}(\{\mathbf{x}, \mathbf{y}^{f,c}\}))],$$

where D_{cmp} is functional on mapping pairs of fashion items into a common space, encoding composed outfits into embeddings, and it pushes $\{\mathbf{x}, \mathbf{y}^{r,c}\}$ far from both $\{\mathbf{x}, \mathbf{y}^{r,!c}\}$ and $\{\mathbf{x}, \mathbf{y}^{f,c}\}$ in the space. For training the generator G, the compatibility loss is defined as follows:

$$\mathcal{L}_{cmp} =$$
$$[\mathbb{E}_{\mathbf{z}_i \sim \mathcal{Z}, \mathbf{x} \sim \mathcal{X}}(D_{cmp}(\{\mathbf{x}, \mathbf{y}^{f,c}\}))$$
$$-\mathbb{E}_{\mathbf{z}_i \sim \mathcal{Z}, \mathbf{x} \sim \mathcal{X}}(D_{cmp}(\{\mathbf{x}, \mathbf{y}^{r,c}\}))] \tag{7.4}$$
$$+[\mathbb{E}_{\mathbf{z}_i \sim \mathcal{Z}, \mathbf{x} \sim \mathcal{X}}(D_{cmp}(\{\mathbf{x}, \mathbf{y}^{f,c}\}))$$
$$-\mathbb{E}_{\mathbf{z}_i \sim \mathcal{Z}, \mathbf{x} \sim \mathcal{X}}(D_{cmp}(\{\mathbf{x}, \mathbf{y}^{f,!c}\}))],$$

where D_{cmp} draws $\{\mathbf{x}, \mathbf{y}^{f,c}\}$ close to $\{\mathbf{x}, \mathbf{y}^{r,c}\}$ and pushes $\{\mathbf{x}, \mathbf{y}^{f,c}\}$ away from $\{\mathbf{x}, \mathbf{y}^{f,!c}\}$.

Diversity Objective. To further enable the generator G to produce diverse clothing images $\{\mathbf{y}_i\}_{i=1}^{n}$ belonging to $\mathcal{Y}$, we explicitly regularize G with the following diversity loss, in line with [5]:

$$\mathcal{L}_{div} = -\mathbb{E}_{\mathbf{y}_i \neq \mathbf{y}_j}[d(\mathbf{y}_i, \mathbf{y}_j)], \tag{7.5}$$

where $d(\cdot, \cdot)$ is a distance function. In our implementation, we used the LPIPS [47] as our distance metric, which can guarantee convergence during the optimization process. In fact, we have also conducted experiments with the L1/L2 distance, as alternatives to LPIPS. However, according to our experiments, it is hard for the alternative models to converge. The underlying reason behind this phenomenon may be ascribed to the fact that the last layer of the pre-trained model has no activation function and the output of the last layer resides in a large range.

Full Objective. For training the generator of our BC GAN, the full objective functions can be summarized as follows:

$$\mathcal{L}_{total} = \mathcal{L}_{adv} + \lambda_1 \mathcal{L}_{div} + \lambda_2 \mathcal{L}_{cmp}, \tag{7.6}$$

where $\mathcal{L}_{adv}$ is a GAN loss, which is responsible for the visual authenticity of the synthesized results; $\mathcal{L}_{div}$ is a diversity loss, which is responsible for synthesizing diverse results; and $\mathcal{L}_{cmp}$ is a compatibility loss, which is responsible for the fashion compatibility of the synthesized results. Moreover, λ_1 and λ_2 are parameters to balance the relative contributions of the loss terms.

Algorithm 7.1 Adversarial training algorithm for BC-GAN

Input: Compatible and diverse outfit pairs $\{\mathbf{x}, \mathbf{y}\} \in \mathcal{X} \times \mathcal{Y}$.
Output: Generator G of BC-GAN.
Pre-train StyleGAN on target domain $\mathcal{Y}$ of our training set; # To obtain f and g
Initialize the parameters of e, D and D_{cmp}, copy the parameters of f and g from the pre-trained
  StyleGAN and freeze them;
for $iter \leftarrow 1$ **to** N_{iter} **do**
  | Sample a batch of $\{\mathbf{x}, \mathbf{y}\}$ from training set;
  | Sample a batch of random latent codes $\{\mathbf{z}_i\}_{i=1}^n$ from $\mathcal{N}(0, \mathrm{I})$;
  | # To train the style embedding discriminator D
  | $\mathcal{L}_{dis} = \mathbb{E}_{\mathbf{z}_i \sim \mathcal{Z}, \mathbf{x} \sim \mathcal{X}}[\log D_{se}(f(\mathbf{z}_i)) + \log(1 - D_{se}(e(\mathbf{x} \oplus f \circ g(\mathbf{z}_i))))]$; # See Eq. (7.1)
  | update parameters $\theta_{D_{se}}$ of D_{se} with $\theta_{D_{se}} \leftarrow \theta_{D_{se}} - \eta \nabla_{\theta_{D_{se}}} \mathcal{L}_{dis}$; # η is a learning rate
  | # To train the compatibility discriminator D_{cmp}
  | $\mathbf{y}_i \leftarrow e \circ g(\mathbf{x} \oplus f \circ g(\mathbf{z}_i))$, where $i \in \{1, \ldots, n\}$;
  | Assign sample $\mathbf{x}$ with samples $\mathbf{y}_i$ ($i \in \{1, \ldots, n\}$), $\mathbf{y}$ into $\{\mathbf{x}, \mathbf{y}^{r,c}\}$, $\{\mathbf{x}, \mathbf{y}^{f,c}\}$, and $\{\mathbf{x}, \mathbf{y}^{r,!c}\}$ pairs;
  | $\mathcal{L}_{cmp\_dis} \leftarrow [\mathbb{E}_{\mathbf{z}_i \sim \mathcal{Z}, \mathbf{x} \sim \mathcal{X}}(D_{cmp}(\{\mathbf{x}, \mathbf{y}^{r,c}\})) - \mathbb{E}_{\mathbf{z}_i \sim \mathcal{Z}, \mathbf{x} \sim \mathcal{X}}(D_{cmp}(\{\mathbf{x}, \mathbf{y}^{r,!c}\}))] +$
  | $\quad [\mathbb{E}_{\mathbf{z}_i \sim \mathcal{Z}, \mathbf{x} \sim \mathcal{X}}(D_{cmp}(\{\mathbf{x}, \mathbf{y}^{r,c}\})) - \mathbb{E}_{\mathbf{z}_i \sim \mathcal{Z}, \mathbf{x} \sim \mathcal{X}}(D_{cmp}(\{\mathbf{x}, \mathbf{y}^{f,c}\}))]$; # See Eq. (7.3)
  | update parameters $\theta_{D_{cmp}}$ of D_{cmp} with $\theta_{D_{cmp}} \leftarrow \theta_{D_{cmp}} - \eta \nabla_{\theta_{D_{cmp}}} \mathcal{L}_{cmp\_dis}$;
  | # To train the generator G
  | $\mathbf{w}_i^{orig} \leftarrow f(\mathbf{z}_i)$, where $i \in \{1, \ldots, n\}$;
  | $\mathbf{w}_i \leftarrow e(\mathbf{x} \oplus g(\mathbf{w}_i^{orig}))$, where $i \in \{1, \ldots, n\}$;
  | $\mathbf{y}_i \leftarrow e \circ g(\mathbf{x} \oplus g(\mathbf{w}_i))$, where $i \in \{1, \ldots, n\}$;
  | Assign sample $\mathbf{x}$ with samples $\mathbf{y}_i$ ($i \in \{1, \ldots, n\}$), $\mathbf{y}$ into $\{\mathbf{x}, \mathbf{y}^{r,c}\}$, $\{\mathbf{x}, \mathbf{y}^{f,c}\}$, and $\{\mathbf{x}, \mathbf{y}^{f,!c}\}$ pairs;
  | $\mathcal{L}_{adv} \leftarrow \mathbb{E}_{\mathbf{z}_i \sim \mathcal{Z}, \mathbf{x} \sim \mathcal{X}}[\log(D_{se}(\mathbf{y}_i)]$; # See Eq. (7.2)
  | $\mathcal{L}_{cmp} \leftarrow [\mathbb{E}_{\mathbf{z}_i \sim \mathcal{Z}, \mathbf{x} \sim \mathcal{X}}(D_{cmp}(\{\mathbf{x}, \mathbf{y}^{f,c}\})) - \mathbb{E}_{\mathbf{z}_i \sim \mathcal{Z}, \mathbf{x} \sim \mathcal{X}}(D_{cmp}(\{\mathbf{x}, \mathbf{y}^{r,c}\}))] +$
  | $\quad [\mathbb{E}_{\mathbf{z}_i \sim \mathcal{Z}, \mathbf{x} \sim \mathcal{X}}(D_{cmp}(\{\mathbf{x}, \mathbf{y}^{f,c}\})) - \mathbb{E}_{\mathbf{z}_i \sim \mathcal{Z}, \mathbf{x} \sim \mathcal{X}}(D_{cmp}(\{\mathbf{x}, \mathbf{y}^{r,!c}\}))]$; # See Eq. (7.4)
  | $\mathcal{L}_{div} \leftarrow -\mathbb{E}_{\mathbf{y}_i \neq \mathbf{y}_j}[d(\mathbf{y}_i, \mathbf{y}_j)]$; # See Eq. (7.5)
  | $\mathcal{L}_{total} \leftarrow \mathcal{L}_{adv} + \lambda_1 \mathcal{L}_{div} + \lambda_2 \mathcal{L}_{cmp}$; # See Eq. (7.6)
  | update parameters θ_G of G with $\theta_G \leftarrow \theta_G - \eta \nabla_{\theta_G} \mathcal{L}_{total}$;
end
return G^*; # Optimized G (includes f, e and g)

7.3.5 The Adversarial Training Process

In this subsection, we present the design of an adversarial training scheme which is used to optimize the generator G of BC-GAN. For clarity, the entire training process of BC-GAN is summarized in Algorithm 7.1. In the beginning, we pre-trained a StyleGAN on the target domain using our training set (shown in line 1). Then the parameters of e, D_{se} and D_{cmp} are initialized, and the parameters of f and g are copied from the above pre-trained StyleGAN and frozen (shown in line 2). The subsequent training process is carried out by applying a gradient descent step on D_{se}, D_{cmp} and G in alternating steps, and using the gradient descent method to update the parameters $\theta_{D_{se}}$, $\theta_{D_{cmp}}$ and θ_G of D_{se}, D_{cmp} and G with a learning rate η, respectively (shown in lines 3–24). In particular, given a batch of pairs of visually-collocated items of clothing $\{\mathbf{x}, \mathbf{y}\}$ and a batch of sampled latent codes

$\{z\}_{i=1}^{n}$ from $\mathcal{N}(0, I)$ (shown in lines 4–5), in the following training steps, we first train our style embedding discriminator D_{se} (shown in lines 6–8). The style embeddings $\{w_i^{orig}\}_{i=1}^{n}$ from f and $\{w_i\}_{i=1}^{n}$ from e are generated and then fed into D_{se} to update its parameters (shown in lines 7–8). Subsequently, the compatibility discriminator D_{cmp} is trained from a contrastive learning perspective (shown in lines 9–13). Different pairs are synthesized and designated by specified labels (shown in lines 10–11). These pairs are fed into D_{cmp} to update its parameters (shown in lines 12–13). At last, the adversarial loss $\mathcal{L}_{adv}$ (shown in line 19), compatibility loss $\mathcal{L}_{cmp}$ (shown in line 20), and diversity loss $\mathcal{L}_{div}$ (shown in line 21) are calculated. Moreover, our generator G is trained with $\mathcal{L}_{total}$ (shown in lines 22–23). When the training process converges, our framework returns the optimized generator G of BC-GAN to synthesize compatible and diverse clothing images.

7.4   Experiments

In this section, the construction of our dataset, named DiverseOutfits, is first introduced. Then we experimentally show the superiority of our proposed method over existing methods in terms of diversity, visual authenticity, and fashion compatibility. Furthermore, we provide an ablation study to verify the effectiveness of the main modules in BC-GAN. Finally, an additional study about continuous translation on the target domain is conducted.

7.4.1   Dataset

When carrying out collocated clothing generation, a large-scale fashion dataset which can provide valid samples for model training and test becomes essential. In fact, several fashion datasets have been constructed for different research purposes, such as UT Zappos50K [44], the Maryland Polyvore dataset [11], and the OutfitSet dataset [50]. However, most of the existing released datasets lack annotations of diverse matching pairs of fashion items. Sometimes they treat the same clothing as different entities. In order to address the issues present in the existing datasets, we have collected a comprehensive dataset, named DiverseOutfit, from a fashion matching community, Polyvore.com. This dataset includes 31,631 outfits, each of which is composed of an upper and lower clothing item. Each outfit has been meticulously curated by fashion experts to ensure compatibility. In order to merge identical entities, for each category of fashion items, LPIPS was employed to gauge the visual similarity of two images of upper (or lower) clothing. If the similarity between these two items of clothing is beyond the set cut-off, they were classified as belonging to the same entity. As shown in Fig. 7.4a, we can observe that each fashion item can match with multiple complementary fashion items. To further verify our observations, we conducted statistics on the distributions in terms of an upper (or lower) clothing item matching how many lower (or upper) clothing. The statistics illustrated in Fig. 7.4b and c show that 58.0% (or 66.6%) of upper (or lower)

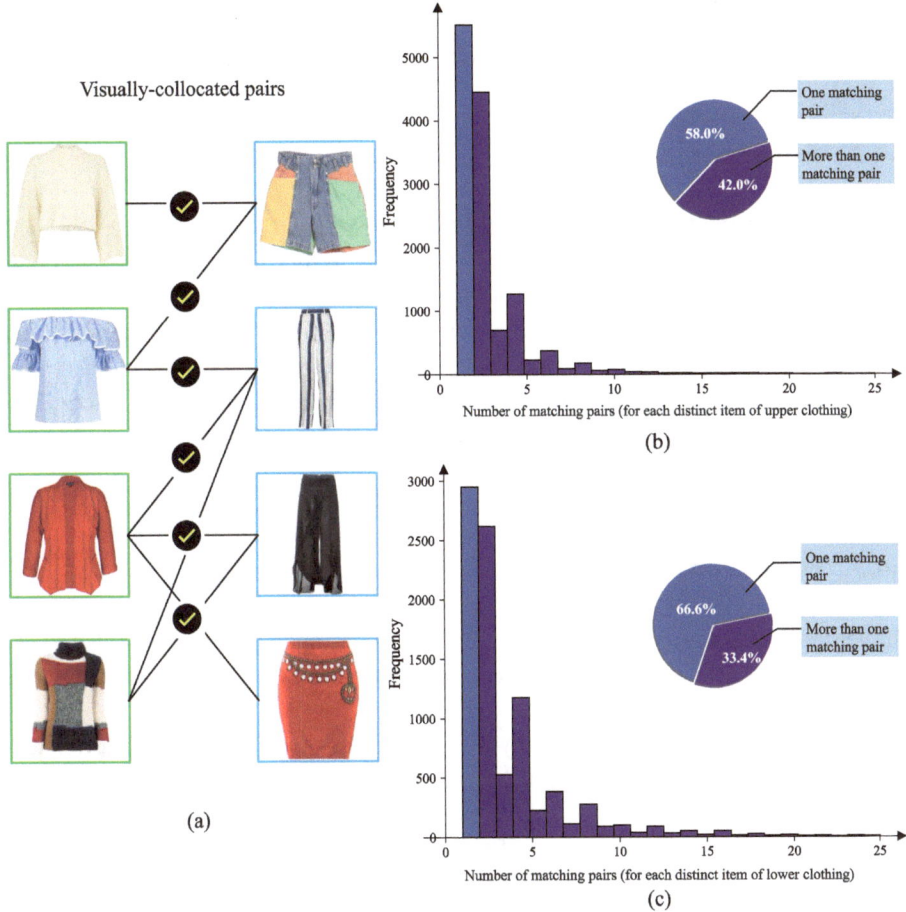

Fig. 7.4 Statistics of our DiverseOutfits dataset: **a** an illustration of multiple possible matching outfits, **b** statistics of matching pairs (for each distinct item of upper clothing), and **c** statistics of matching pairs (for each distinct item of lower clothing)

clothing items have one matching fashion item, while 42.0% (or 33.4%) of upper (or lower) clothing have more than one matching fashion item, indicating the diverse compatible rules hidden in the DiverseOutfits dataset. We then partitioned these outfits into two subsets to form a training set and a test set. In fact, as a multimodal I2I translation method, our task has two different settings: the 'upper → lower' and 'lower → upper' directions. For the 'upper → lower' setting, we split the outfit composition pairs into training and test sets randomly. There is no overlap for the upper clothing between the training and test sets. The training and test sets have a ratio of 4 : 1, according to the number of matching pairs. For the 'upper → lower', the training set has 25,305 matching pairs, and the test set has 6,326 matching pairs. Similarly, we constructed the dataset for the 'lower → upper' setting, in which its training and test sets have 25,305 and 6,326 matching pairs, respectively.

7.4.2   Experimental Settings

In our experiments, all images were resized to 256×256 and L was set to 512. During the training phase, the batch size of our BC-GAN was set to four, and the number of training iterations was set to 50,000. All experiments were performed on a single A6000 graphics card, and the implementation was carried out in PyTorch [30]. BC-GAN was trained with an Adam [20] optimizer with $\beta_1 = 0$ and $\beta_2 = 0.99$, and the learning rate η was set to 2×10^{-4}. In order to make the adversarial training more stable, the style embedding discriminator pool was set with a size of 100 in our implementation. The coefficients of Eq. (7.6) were empirically set with $\lambda_1 = 1$ and $\lambda_2 = 3$.

7.4.3   Evaluation Metrics

The generated clothing images were evaluated from three important aspects: diversity, visual authenticity, and fashion compatibility. For clarity, the evaluation metrics used are described as follows:

Diversity Measurement. A diversity measurement is used to measure the diversity of the synthesized images $\{\mathbf{y}_i\}_{i=1}^n$ lying in the target domain $\mathcal{Y}$ for the same input image $\mathbf{x}$ from the source domain $\mathcal{X}$. Following previous studies [5, 15], we use LPIPS as our evaluation metric. It calculates the mean distance of each possible combination of synthesized images as follows:

$$\text{LPIPS} = \mathbb{E}_{\mathbf{x} \sim \mathcal{X}} [\sum_{i=1}^{N} \sum_{j=i+1}^{N} \frac{2 \times d(\mathbf{y}_i, \mathbf{y}_j)}{N \times (N-1)}], \tag{7.7}$$

where N is the number of synthesized images for each input image, and we set $N = 10$. The distance function $d(\cdot, \cdot)$ was implemented by LPIPS with a pre-trained AlexNet[1] to evaluate the performance in terms of diversity. Here, a higher score indicates better diversity.

Visual Authenticity Measurement. Previous studies [5, 28, 45, 48, 50] have suggested that the Fréchet inception distance [12] (FID) can be used to estimate the visual authenticity of synthesized images in a high-level feature space mapped by a pre-trained model. In particular, FID measures the similarity of the statistics of the features of synthesized and real images. We used the implementation[2] of FID to measure the visual authenticity of synthesized images. A lower FID value means better visual authenticity.

---

[1] https://github.com/richzhang/PerceptualSimilarity.
[2] https://github.com/mseitzer/pytorch-fid.

Fashion Compatibility Measurement. A fashion compatibility measurement is used to gauge the visually-collocated degree of a given clothing image and synthesized clothing image. Following Zhou et al. [48], we adopt fashion fill-in-the-blank best times (F^2BT) as a metric to estimate the fashion compatibility of synthesized fashion items with a given item. To make this chapter self-contained, a brief introduction of F^2BT is given here. F^2BT adopts a pre-trained fashion compatibility predictor φ in an open-source toolbox MMFashion.[3] It should be noted that the dataset on which φ was trained, Maryland Polyvore dataset [11], is different from that our experiments. φ plays a role in comparing two composed outfits as well as predicting which of them is better. We count the number of times that each method beats all other methods, which is under the concept of 'one-vs-rest'. Formally, it is given as follows:

$$F^2BT(m_i) = \mathbb{E}_{\mathbf{o}_k}[\frac{\prod_{j=1, j \neq i}^{N_m} \mathbb{1}(\varphi(\mathbf{o}_{i,k}) > \varphi(\mathbf{o}_{j,k}))}{N_m}], \qquad (7.8)$$

where $\mathbb{1}(\cdot)$ is a conditional function; for the i-th method in the compared set, we count the average number of times that this method beats the other methods in terms of fashion compatibility; $\mathbf{o}_{i,k}$ is the k-th outfit synthesized by the i-th method m_i, and $\mathbf{o}_{j,k}$ has a similar definition. It should be noted that all values are normalized by the size of the set of compared methods. Here, a higher value means better fashion compatibility.

7.4.4 On Performance Comparison

Compared Methods. Since there are few other models that can carry out the task considered here, our proposed BC-GAN is evaluated against five state-of-the-art methods, which are closely related to our method. The five compared baseline methods include MUNIT [15], DRIT [22], DRIT++ [23], StarGAN-v2 [5], and SAVI2I [28]. We give a brief introduction to these baselines as follows:

- **MUNIT** [15] attempts to disentangle images into content codes and style codes in a high-level feature space with two different encoders. To synthesize diverse images, it combines content codes and style codes encoded from different images. During the model training, the reconstruction loss is a key component to learn conditional mapping.
- **DRIT** [22] proposes an approach based on disentangled representation for producing diverse outputs in I2I translation. It attempts to disentangle the information of an image into domain-invariant content space and domain-specific attribute space.
- **DRIT++** [23] is an improved version of DRIT. It extends the original framework from two domains to multiple domains for I2I translation. Moreover, a new mode-seeking regularization is proposed to improve the diversity of the synthesized results.

---

[3] https://github.com/open-mmlab/mmfashion.

Table 7.1 Comparison of BC-GAN and baselines on upper → lower setting in terms of diversity (LPIPS), visual authenticity (FID), and fashion compatibility (F^{2}BT) (here, for all metrics except FID, higher is better)

Method	Evaluation metrics on upper → lower		
	LPIPS(↑)	FID(↓)	F^{2}BT(↑)[a] (%)
MUNIT [15]	0.479	60.016	16.5
DRIT [22]	0.248	93.359	14.2
DRIT++ [23]	0.304	74.033	14.0
StarGAN-v2 [5]	0.343	92.489	13.6
SAVI2I [28]	0.397	123.944	20.0
BC-GAN (ours)	**0.504**	**52.370**	**22.9**

[a]*Note* Percentages may not sum to 100 because of rounding

- **StarGAN-v2** [5] is an improved version of StarGAN [4]. Compared with StarGAN, which only supports unimodal I2I translation, StarGAN-v2 supports multimodal translation.
- **SAVI2I** [28] is the first framework supporting continuous multimodal I2I translation. It employs signed attribute vectors to control the continuous I2I translation.

For a fair comparison, all model implementations were based on the original codes reported by the authors. For each translation direction in diverse collocated clothing synthesis, these baseline models were trained from scratch. For instance, in the 'upper → lower' translation, the source and target domains are upper and lower clothing, respectively.

Comparison of Results. For comparison, the quantitative evaluation results of all compared methods are given in Tables 7.1 and 7.2. Moreover, a visual comparison of synthesized results is given in Fig. 7.5. As described in Sect. 7.4.1, our problem has two translation directions: 'upper → lower' and 'lower → upper' settings. For the setting of 'upper → lower' translation, the results of the comparison of BC-GAN and other baselines are given in Table 7.1. It shows that our proposed BC-GAN consistently outperforms other multimodal I2I translation methods in terms of diversity (LPIPS), visual authenticity (FID), and fashion compatibility (F^{2}BT). The results of 'upper → lower' and 'lower → upper' directions are analyzed independently as follows:

- **The 'upper → lower' direction**: (1) For the metric of diversity, our method improves by approximately 0.025 in comparison with the second-best method. From the synthesized results in Fig. 7.5a, it is clear that our model produces the most diverse results, especially for shape and texture. Taking the second and third samples for the 'upper → lower' setting as an example, we can observe that BC-GAN synthesizes shorts, skirts, or trousers with diverse shapes and textures. Compared with our method, the results of other baselines are not as promising as ours in terms of diversity. In particular, the other

Table 7.2 Comparison of BC-GAN and baselines on lower → upper setting in terms of diversity (LPIPS), visual authenticity (FID), and fashion compatibility (F²BT) (here, for all metrics except FID, higher is better)

Method	Evaluation metrics on lower → upper		
	LPIPS(↑)	FID(↓)	F²BT(↑)[a] (%)
MUNIT [15]	0.547	47.036	16.6
DRIT [22]	0.263	77.858	16.1
DRIT++ [23]	0.335	61.170	14.4
StarGAN-v2 [5]	0.386	42.626	14.7
SAVI2I [28]	0.378	118.546	15.4
BC-GAN (ours)	**0.605**	**38.981**	**22.8**

[a]*Note* Percentages may not sum to 100 because of rounding

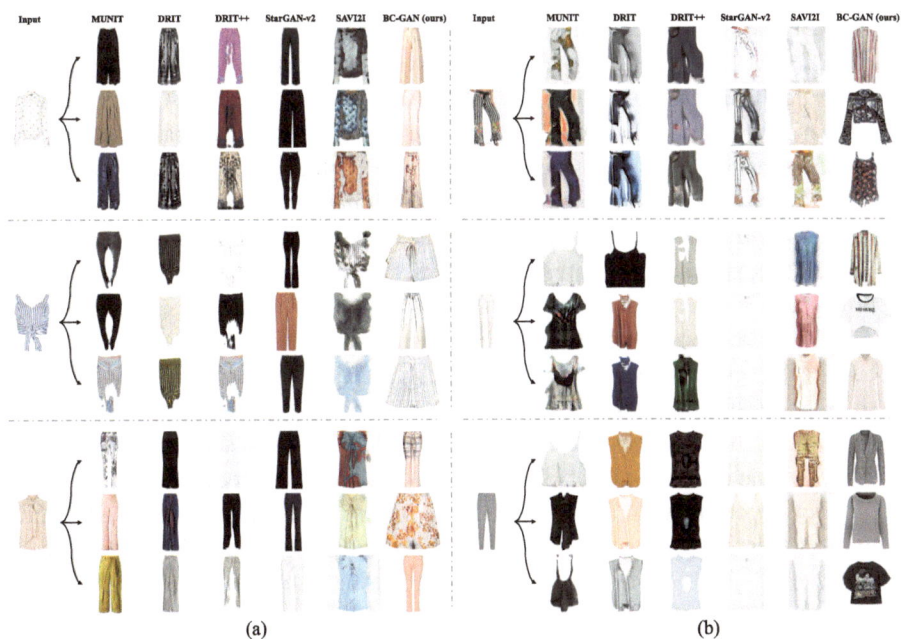

(a)   (b)

Fig. 7.5 Comparisons between our BC-GAN and other multimodal I2I translation methods, which are MUNIT [15], DRIT [22], DRIT++ [23], StarGAN-v2 [5], and SAVI2I [28], in terms of **a** translation on upper → lower setting, and **b** translation on lower → upper setting

baselines, except for SAVI2I, cannot produce images of shorts or skirts. The StarGAN-v2 method can only synthesize trouser images and lacks shape diversity even for different inputs. On the contrary, SAVI2I produces results that are similar to its inputs. This may be ascribed to the fact that SAVI2I is more focused on continuous I2I translation so that the result of the synthesis can only exert influence over the input in color space.

(2) For the metric of visual authenticity, Table 7.1 shows the FID of BC-GAN and other baselines on the 'upper → lower' setting. We observe that our BC-GAN achieves the best score of visual authenticity in terms of FID. It outperforms the second-best method by a substantial margin: approximately 7.646. Among the baselines compared, MUNIT demonstrated the best performance in terms of FID. This can be attributed to its unique approach by employing separate style and content encoders, resulting in increased network capacity when compared to other alternatives. For qualitative observation, Fig. 7.5a gives an illustration of synthesized images. It is clear that our method produces the most photo-realistic images in comparison to other methods. (3) For the metric of fashion compatibility, to reduce the subjectivity of the evaluation, F^2BT is adopted as a metric to measure the results of BC-GAN and the baselines. As shown in Table 7.1, BC-GAN delivers the best fashion compatibility score in terms of F^2BT. Specifically, it achieves the best score with an advantage of 2.9% compared with the second-best method. In Fig. 7.5a, the synthesized results of our model possess elements that harmonize with the given inputs. This indicates that our method synthesizes better results in terms of fashion compatibility. Meanwhile, although SAVI2I achieves the second-best fashion compatibility metric as shown in Table 7.1, its synthesized images convey patterns that are too similar to its input as shown in Fig. 7.5a. On the contrary, BC-GAN has better texture generation, as its results harmonize more with their inputs, from the perspective of human observation. Taking the second sample for the 'upper → lower' setting, we observe that BC-GAN produces lower clothing with vertical stripes which are compatible with the fashion style of the given upper clothing.

- **The 'lower → upper' direction**: (1) For the metric of diversity, BC-GAN achieves the best LPIPS and has an improvement of approximately 0.058 in comparison with the second-best method as shown in Table 7.2. Figure 7.5b also indicates that our method produces the best diversity among all methods. Taking the first sample for the 'lower → upper' setting as an example, we can observe that BC-GAN produces the most different types of upper clothing. (2) For the metric of visual authenticity, BC-GAN achieves the best score in terms of FID and improves by 3.645 in comparison with the second-best method, as shown in Table 7.2. Considering the visual comparison, the generated results listed in Fig. 7.5b exhibit that BC-GAN synthesizes the most photo-realistic results among all methods. (3) For the fashion compatibility measurement, Table 7.2 illustrates that BC-GAN delivers the highest F^2BT score among all methods. Moreover, Fig. 7.5b also confirms that our method synthesizes the results with higher fashion compatibility. For example, it is observed that the synthetic clothing items are, with pink elements, compatible with the given clothing in the first sample for the 'lower → upper' setting.

User Study: In addition to the quantitative assessment with the above metrics, we undertook a user study for BC-GAN to delve deeper into the perceptual evaluation of the fashion compatibility exhibited by the generated clothing images listed in Table 7.3. From the DiverseOutfits dataset, we randomly selected 100 images for both the 'upper → lower' and

Table 7.3 User study results: comparative preference rate of BC-GAN (ours) and baselines (Here, 'Real' and 'Random' refer to ground truth and randomly composed outfits from DiverseOutfits dataset, respectively)

	BC-GAN > MUNIT	BC-GAN > DRIT	BC-GAN > DRIT++	BC-GAN > StarGAN-v2	BC-GAN > SAVI2I	BC-GAN > Real	BC-GAN > Random
Preference (%)	85.6	88.8	91.8	78.6	95.5	40.7	76.7

'lower → upper' translation directions as given clothing. Subsequently, we performed a user study to validate the superiority of our method. Our user study involved 10 participants who were not affiliated with the authors, yielding a total of 7,000 votes. Each participant was presented with three images: the original clothing, the clothing synthesized through our proposed method, and the clothing synthesized either by a comparable baseline or from the DiverseOutfits dataset. They were instructed to select the outfit that displayed superior collocation. The results of the user study shown in Table 7.3 demonstrate the substantial superiority of our method over the baseline approaches. It also indicates that the compatibility of fashion items synthesized by our method closely parallels the matching degree of expert-constructed outfits in the DiverseOutfits dataset.

7.4.5 On Comparison with Baselines in Collocated Clothing Synthesis

Since diversified collocated clothing synthesis is a new task proposed by this research, to further validate the effectiveness of our proposed model, we conducted a comparison with other collocated clothing synthesis baselines: Attribute-GAN [25], CA-GAN [26], and COutfitGAN [48]. The implementations are all based on the official codes from the original authors of these works. The evaluated metrics are shown in Tables 7.4 and 7.5. It should be noted here that these baselines in collocated clothing synthesis only support generating one candidate of clothing rather than multiple candidates of clothing. Consequently, the metric of diversity (LPIPS) cannot be performed on these baselines. For the measurement, it is clear that our method has the best visual authenticity in terms of FID. Moreover, it was observed that the metrics of Attribute-GAN and CA-GAN were subpar. Upon analysis, we postulated that this could be attributed to the integration of attribute information of clothing, which seemed to render the training process less robust and stable. For the measurement of fashion compatibility, BC-GAN achieves the highest value of F^2BT among these baselines in collocated clothing synthesis.

Table 7.4 Comparison of BC-GAN and baselines on upper → lower setting in terms of diversity (LPIPS), visual authenticity (FID), and fashion compatibility (F^{2}BT) (here, for all metrics except FID, higher is better)

Method	Evaluation metrics on upper → lower		
	LPIPS (↑)	FID (↓)	F^{2}BT(↑[a]) (%)
Attribute-GAN [25]	–	165.449	26.4
CA-GAN [26]	–	212.596	14.2
COutfitGAN [48]	–	90.608	29.2
BC-GAN (ours)	**0.504**	**52.370**	**30.2**

[a]*Note* Percentages may not sum to 100 because of rounding

Table 7.5 Comparison of BC-GAN and baselines on lower → upper setting in terms of diversity (LPIPS), visual authenticity (FID), and fashion compatibility (F^{2}BT) (here, for all metrics except FID, higher is better)

Method	Evaluation metrics on lower → upper		
	LPIPS(↑)	FID(↓)	F^{2}BT(↑[a]) (%)
Attribute-GAN [25]	–	233.905	26.2
CA-GAN [26]	–	159.600	31.6
COutfitGAN [48]	–	67.787	7.0
BC-GAN (ours)	**0.605**	**38.981**	**35.2**

[a]*Note* Percentages may not sum to 100 because of rounding

7.4.6   On Ablation Study

This subsection presents the results of three sets of ablation studies carried out to validate the effectiveness of diversity loss, style embedding discriminator, and fashion compatibility discriminator.

Effectiveness of Diversity Loss: The diversity loss is used to improve the diversity of synthetic results of BC-GAN. To investigate the effectiveness of diversity loss, we explore its impact on quantitative performance. The comparative results are given in Table 7.6. The results show that BC-GAN with diversity loss outperforms BC-GAN without diversity loss by a substantial margin: approximately 0.086 (or 0.078) on the 'upper → lower' (or 'lower → upper') translation setting. For qualitative comparison, we also provided certain synthetic samples in Fig. 7.6. Taking the first and second columns of Fig. 7.6 as examples, we can see that BC-GAN without $\mathcal{L}_{div}$ only produces images of trousers for their given inputs. Moreover, the third and fourth columns of Fig. 7.6 show that the synthetic results deliver very similar colors. We believe this is caused by the fact that removing $\mathcal{L}_{div}$ degrades the variance of the synthetic samples in the target domain. This verifies that the addition of diversity loss benefits diversity in terms of both quantitative and qualitative evaluations.

Table 7.6 Comparison of BC-GAN with other variants without diversity loss in terms of diversity (LPIPS)

Method	Setting	LPIPS(↑)
BC-GAN w/o $\mathcal{L}_{div}$	upper → lower	0.418
	lower → upper	0.527
BC-GAN (ours)	upper → lower	**0.504**
	lower → upper	**0.605**

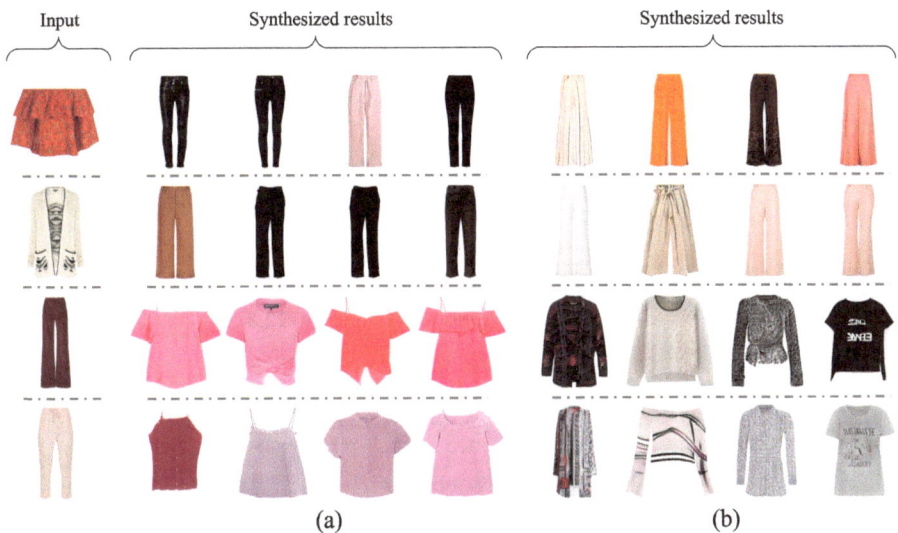

Fig. 7.6 Visual comparison results for the effectiveness of diversity loss: **a** BC-GAN without diversity loss, and **b** BC-GAN with diversity loss (ours)

Effectiveness of Style Embedding Discriminator: We evaluated the effectiveness of the style embedding discriminator D_{se} from two perspectives. Initially, we trained our BC-GAN without incorporating the D_{se}. As shown in Table 7.7, the BC-GAN, lacking the style embedding discriminator, fails to ensure the realism of the synthesized clothing images. Subsequently, we trained our BC-GAN using a pixel-level discriminator instead of one in the $\mathcal{W}$ space. In Table 7.7, 'BC-GAN w/ D_{pix}' implies that the BC-GAN was trained with the original StyleGAN's real/fake discriminator, rather than the D_{se}. The results reveal that the BC-GAN with D_{pix} is incapable of generating photo-realistic images, as evidenced by the high FID. This finding may seem counter-intuitive. However, upon analysis, it was deduced that the D_{pix} supervises the generator's training by acting on the generated image, but its supervisory effect struggles to propagate to the encoding network e over a long distance. Specifically, its pixel-level supervision is applied to the image generated by the synthesis network g. But, as the weights of the synthesis network are frozen during the

Table 7.7 Comparison of BC-GAN with other variants without style embedding discriminator in terms of visual authenticity (FID)

Method	Setting	FID($\downarrow$)
BC-GAN w/o D_{se}	upper $\rightarrow$ lower	339.445
	lower $\rightarrow$ upper	331.327
BC-GAN w/ D_{pix}	upper $\rightarrow$ lower	291.862
	lower $\rightarrow$ upper	260.163
BC-GAN (ours)	upper $\rightarrow$ lower	**52.370**
	lower $\rightarrow$ upper	**38.981**

Table 7.8 Comparison of BC-GAN with other variants without compatibility discriminator in terms of fashion compatibility (F^2BT)

Method	Setting	F^2BT ($\uparrow$) (%)
BC-GAN w/o D_{cmp}	upper $\rightarrow$ lower	14.9
	lower $\rightarrow$ upper	17.0
BC-GAN (ours)	upper $\rightarrow$ lower	**22.9**
	lower $\rightarrow$ upper	**22.8**

BC-GAN training process, when its influence is back-propagated to the encoding network e, the weight update is hard to be achieved. In contrast, the style embedding discriminator D_{se} proposed in this research is designed to directly control the latent codes encoded from the encoding network e. Once the latent codes align with the distribution of the $\mathcal{W}$ space, the generated image can maintain a high level of visual fidelity.

Effectiveness of Compatibility Discriminator: The compatibility discriminator is designed in a contrastive learning perspective to further improve the fashion compatibility in comparison to that of previous studies [25, 26, 45]. To investigate the effectiveness of our compatibility discriminator, we validate it from two aspects. For the first aspect, we first trained BC-GAN without using our compatibility discriminator. Then we compared this version with our original BC-GAN in terms of the aforementioned fashion compatibility metric, F^2BT. As shown in Table 7.8, it is clear that our compatibility discriminator plays an important role in providing visually-collocated supervision on training BC-GAN by showing its contributions to the improvement in F^2BT. We also generated some samples produced by BC-GAN with or without the use of the compatibility discriminator in Fig. 7.7. Obviously, synthetic clothing items harmonize better, which raises the fashion compatibility with the given clothing. For the second aspect, previous studies on collocated clothing generation only employed a discriminator to supervise the generator on real and fake pairs of given clothing and synthetic clothing. Here, we further compare our contrastive learning-based compatibility discriminator with the previous designs. In Table 7.8, 'BC-GAN w/o

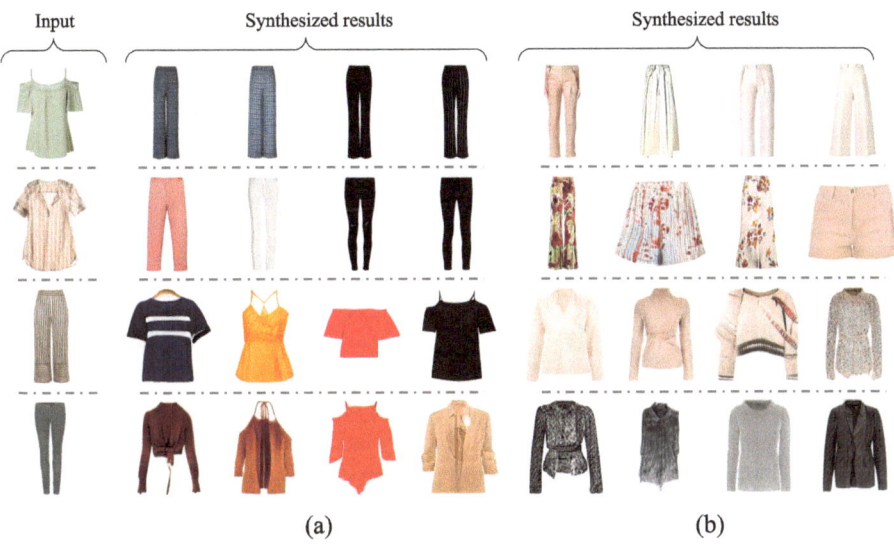

Input   Synthesized results   Synthesized results

(a)   (b)

Fig. 7.7 Visual comparison results for the effectiveness of compatibility discriminator: **a** BC-GAN without compatibility discriminator, and **b** BC-GAN with compatibility discriminator (ours)

Table 7.9 Comparison of BC-GAN with other variants without contrastive learning in terms of fashion compatibility (F^2BT)

Method	Setting	F^2BT (↑) (%)
BC-GAN w/o contrastive learning	upper → lower	15.4
	lower → upper	17.8
BC-GAN (ours)	upper → lower	**22.9**
	lower → upper	**22.8**

contrastive learning' indicates that BC-GAN only takes real and fake pairs as its input. From Table 7.8, it is clear that BC-GAN can enhance its performance in terms of fashion compatibility by a significant amount (Table 7.9).

7.4.7   On Feature Visualization Learned by the Encoding Network e

To examine the impacts of the encoding network e, we visualize the features learned by e with t-distributed stochastic neighbor embedding (t-SNE) [13]. In the BC-GAN proposed in this study, the source domain image $\mathbf{x}$ and the randomly synthesized $\mathbf{y}_i^{orig}$ from the target domain are concatenated and input into the encoding network e to obtain the required latent code $\mathbf{w}_i$, which is subsequently used to generate the final image $\mathbf{y}_i$. In this section, we input each $\mathbf{x}$ and multiple randomly synthesized $\mathbf{y}_i$ into e to obtain $\mathbf{w}_i$ and ultimately use the average feature $\mathbf{w}$ of multiple $\mathbf{w}_i$ for visual analysis with t-SNE. Here, the number of randomly synthesized $\mathbf{y}_i^{orig}$ was set to five. As depicted in Fig. 7.9, it is evident that similar clothing images consistently cluster together. This observation suggests that our model can effectively align randomly synthesized images in the target domain to the style of given clothing $\mathbf{x}$ from the source domain, for both the 'upper $\rightarrow$ lower' and 'lower $\rightarrow$ upper' translations.

7.4.8   On Additional Study

Our BC-GAN is designed with the idea of "encoding random images with the given image first, decoding them for better fashion compatibility later." Attributable to the disentanglement of the learned $\mathcal{W}$ space, we can easily perform continuous translation on the target domain. In this subsection, we conduct an additional study to validate whether BC-GAN works under the above-mentioned mechanism. In previous research [28], continuous I2I translation aims at morphing images from the source domain $\mathcal{X}$ into the target domain $\mathcal{Y}$. However, continuous I2I translation between upper clothing and lower clothing makes no sense in our research problem. In real scenarios, users do not need an image looking both like the upper and the lower clothing. On the contrary, by taking advantage of the disentangled features in the $\mathcal{W}$ space, our framework can support I2I translation from a randomly sampled synthetic clothing item to a new clothing item which is more compatible with a given clothing item from a visual perspective. Specifically, the continuous translation on the target domain is a linear interpolation based on the style embedding. Formally, it can be formulated as follows:

$$\mathbf{w}_i^{mix} = \mathbf{w}_i^{orig} \times (1 - \alpha) + \mathbf{w}_i \times \alpha, \tag{7.9}$$

where $\mathbf{w}_i^{mix}$ is a style embedding mixed from $\mathbf{w}_i^{orig}$ and $\mathbf{w}_i$ with the ratio α. The mixed style embedding $\mathbf{w}_i^{mix}$ is fed into the pre-trained synthesis network g to obtain an RGB image. Here, we compare the fashion compatibilities coming from different mix ratios of random style embeddings and encoded style embeddings. As shown in Table 7.10, we compare different mix ratios in the style embeddings. We adopt five different mix ratios in {0.00, 0.25, 0.50, 0.75, 1.00} to conduct experiments in terms of fashion compatibility. For 0.00 versus 0.25, we count the number of times that the synthetic results with a ratio of 0.25 are better than that of 0.00 in terms of fashion compatibility. Here, the fashion compatibility comparison used the fashion compatibility predictor φ mentioned in Sect. 7.4.3. For example, the first row of Table 7.10 on 'upper $\rightarrow$ lower' setting means the synthesized results using a mix ratio of 0.25 beat the results of the mix ratio of 0.00 over 50.5% of the time. As shown in Table 7.10, we can observe that a higher mix ratio is able to improve the fashion compatibility of the outfits. Meanwhile, we see that 0.00 versus 0.25 on 'lower $\rightarrow$ upper' has a decrease in terms of fashion compatibility. The worse performance may stem from the fact that the conversion from $\mathbf{w}_i^{orig}$ to $\mathbf{w}_i$ cannot be easily disentangled for a high-level learning,

Table 7.10 The times of better fashion compatibility (F^{2}BT) when the mixed style embedding is provided a larger mix ratio

Mix ratio α	Setting	
	Upper $\rightarrow$ lower (%)	Lower $\rightarrow$ upper (%)
0.00 versus 0.25	50.5	49.3
0.25 versus 0.50	53.4	52.0
0.50 versus 0.75	54.4	53.1
0.75 versus 1.00	52.8	51.1

i.e., fashion compatibility learning. Moreover, it is observed that the fashion compatibility of the synthesized samples improves with an increase of mix ratio, as shown in Fig. 7.8. Taking the first given upper clothing as an example, the synthesized results become more and more compatible with an increase of the mix ratio. This indicates that despite taking a random synthesized image to produce diverse images, BC-GAN is able to encode random information into valid samples and improve the fashion compatibility of the synthesized clothing.

Fig. 7.8 Linear interpolation between $\mathbf{w}_i^{orig}$ and $\mathbf{w}_i$, where $\mathbf{w}_i^{orig}$ is a style embedding mapped from a random latent code into $\mathcal{W}$ space, and $\mathbf{w}_i$ is a style embedding encoded from a given clothing image and a random synthesized image into $\mathcal{W}$ space: **a** continuous translation on the lower domain, and **b** continuous translation on the upper domain

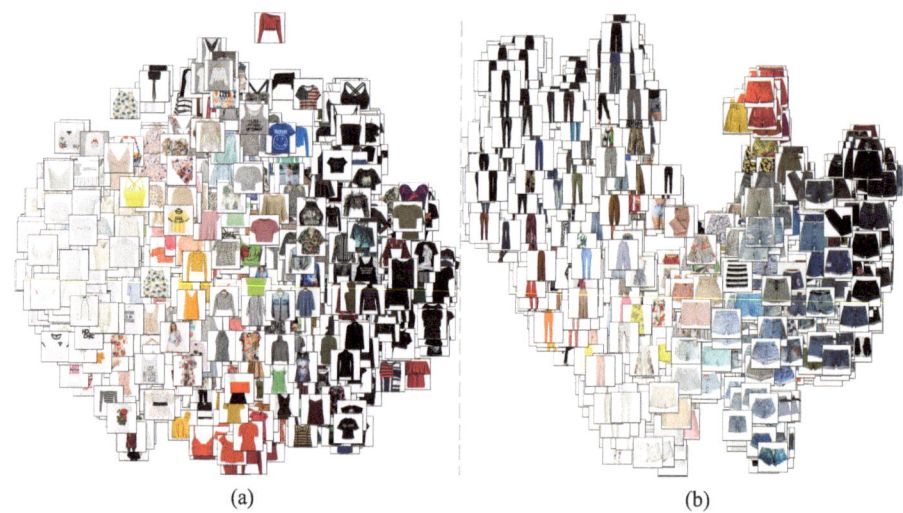

(a)   (b)

Fig. 7.9 Visualization of the features learned by encoding network e via t-SNE: **a** 'upper $\rightarrow$ lower' setting, and **b** 'lower $\rightarrow$ upper' setting. For each direction, a sample of 1,000 images was utilized for visualization

7.4.9   On Sensitivity Study

In order to investigate the effects of the coefficients utilized in the weighting of training losses, we conducted an analysis of the outcomes derived from the BC-GAN model with varying weight parameters, namely λ_1 and λ_2, designated for $\mathcal{L}_{div}$ (diversity loss) and $\mathcal{L}_{cmp}$ (compatibility loss), respectively. A parametric study was carried out specifically for the 'upper $\rightarrow$ lower' translation setting. The results of this investigation are visually represented in Fig. 7.10. Here, the hyper-parameter λ_1 ranges from $\{0, 1, 3\}$, while the hyper-parameter λ_2 comes from $\{0, 1, 3, 5\}$. Figure 7.10a reveals that the LPIPS diversity metric exhibits improvement as the weight assigned to diversity loss, $\mathcal{L}_{div}$, increases. Conversely, an increase in the weight allocated to compatibility loss results in a deterioration of the diversity metric. As shown in Fig. 7.10b, it is evident that the FID metric, indicative of visual authenticity,

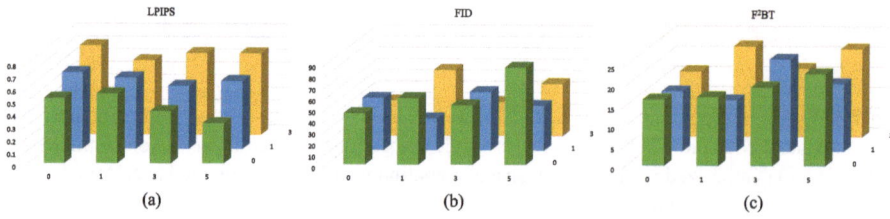

(a)   (b)   (c)

Fig. 7.10 Results of BC-GAN against different settings of weight coefficients of training losses

experiences a decline as both weights increase. This can be attributed to the fact that the augmentation of these weights diminishes the effect of the adversarial loss. Furthermore, Fig. 7.10c demonstrates that the F^2BT metric, a measure of fashion compatibility, improves as the weight for compatibility increases. The findings suggest that a balance needs to optimize the quality of the synthesized images in terms of their diversity, visual authenticity, and fashion compatibility for the task of diverse collocated clothing synthesis. The most effective tradeoff was achieved with the BC-GAN settings of $\lambda_1 = 1$ and $\lambda_2 = 3$.

7.5   Conclusion and Future Work

In this chapter, we proposed a new collocated clothing generation framework by leveraging the power of a pre-trained model. Our framework can synthesize multiple visually-collocated and diverse items of clothing based on the given clothing at the same time. It is composed of a generator to produce targeted images, a style embedding discriminator to ensure the encoded style embedding fits the distribution of latent space of the used pre-trained model, and a compatibility discriminator to supervise the generator capable of synthesizing visually-collocated clothing images. Experiments on a large-scale dataset constructed by ourselves show that the proposed method outperforms alternative methods in terms of diversity, visual authenticity, and fashion compatibility. In future work, it would be interesting to synthesize diversified fashion items in multiple fashion categories in a more controllable manner. In addition, diffusion probabilistic models [14, 32] have been experiencing a surge in interest due to their training stability and their promising results on text-to-image generation. In spite of this, GANs are currently faster in synthesizing images and use less memory when compared to diffusion models during the training phase. In the future, developing collocated clothing synthesis based on advanced diffusion models driven by multi-modal information deserves further investigation.

References

1. Arjovsky, M., Chintala, S., Bottou, L.: Wasserstein generative adversarial networks. In: Proceedings of the International Conference on Machine Learning, pp. 214–223. PMLR (2017)
2. Bau, D., Zhu, J.Y., Wulff, J., Peebles, W., Strobelt, H., Zhou, B., Torralba, A.: Seeing what a gan cannot generate. In: Proceedings of the IEEE/CVF International Conference on Computer Vision, pp. 4502–4511 (2019)
3. Chen, T., Kornblith, S., Norouzi, M., Hinton, G.: A simple framework for contrastive learning of visual representations. In: Proceedings of the International Conference on Machine Learning, pp. 1597–1607. PMLR (2020)
4. Choi, Y., Choi, M., Kim, M., Ha, J.W., Kim, S., Choo, J.: StarGAN: Unified generative adversarial networks for multi-domain image-to-image translation. In: Proceedings of the IEEE/CVF Conference on Computer Vision and Pattern Recognition, pp. 8789–8797 (2018)

5. Choi, Y., Uh, Y., Yoo, J., Ha, J.W.: StarGAN v2: Diverse image synthesis for multiple domains. In: Proceedings of the IEEE/CVF Conference on Computer Vision and Pattern Recognition, pp. 8185–8194 (2020)
6. Chopra, S., Hadsell, R., LeCun, Y.: Learning a similarity metric discriminatively, with application to face verification. In: Proceedings of the IEEE/CVF Conference on Computer Vision and Pattern Recognition, vol. 1, pp. 539–546. IEEE (2005)
7. Cui, Z., Li, Z., Wu, S., Zhang, X.Y., Wang, L.: Dressing as a whole: Outfit compatibility learning based on node-wise graph neural networks. In: Proceedings of the ACM International Conference on World Wide Web Conference, pp. 307–317 (2019)
8. Deng, J., Guo, J., Xue, N., Zafeiriou, S.: Arcface: Additive angular margin loss for deep face recognition. In: Proceedings of the IEEE/CVF Conference on Computer Vision and Pattern Recognition, pp. 4690–4699 (2019)
9. Goodfellow, I., Pouget-Abadie, J., Mirza, M., Xu, B., Warde-Farley, D., Ozair, S., Courville, A., Bengio, Y.: Generative adversarial nets. In: Proceedings of the Advances in Neural Information Processing Systems, vol. 27, pp. 1–9 (2014)
10. Hadsell, R., Chopra, S., LeCun, Y.: Dimensionality reduction by learning an invariant mapping. In: Proceedings of the IEEE/CVF Conference on Computer Vision and Pattern Recognition, vol. 2, pp. 1735–1742. IEEE (2006)
11. Han, X., Wu, Z., Jiang, Y.G., Davis, L.S.: Learning fashion compatibility with bidirectional LSTMs. In: Proceedings of the ACM International Conference on Multimedia, pp. 1078–1086 (2017)
12. Heusel, M., Ramsauer, H., Unterthiner, T., Nessler, B., Hochreiter, S.: GANs trained by a two time-scale update rule converge to a local Nash equilibrium. In: Proceedings of the Advances in Neural Information Processing Systems, pp. 1–12 (2017)
13. Hinton, G.E., Roweis, S.: Stochastic neighbor embedding (2002)
14. Ho, J., Jain, A., Abbeel, P.: Denoising diffusion probabilistic models. Proceedings of the Advances in Neural Information Processing Systems **33**, 6840–6851 (2020)
15. Huang, X., Liu, M.Y., Belongie, S., Kautz, J.: Multimodal unsupervised image-to-image translation. In: Proceedings of the European Conference on Computer Vision, pp. 172–189 (2018)
16. Isola, P., Zhu, J.Y., Zhou, T., Efros, A.A.: Image-to-image translation with conditional adversarial networks. In: Proceedings of the IEEE/CVF Conference on Computer Vision and Pattern Recognition, pp. 1125–1134 (2017)
17. Karras, T., Aittala, M., Hellsten, J., Laine, S., Lehtinen, J., Aila, T.: Training generative adversarial networks with limited data. In: Proceedings of the Advances in Neural Information Processing Systems, pp. 12104–12114 (2020)
18. Karras, T., Laine, S., Aila, T.: A style-based generator architecture for generative adversarial networks. In: Proceedings of the IEEE/CVF Conference on Computer Vision and Pattern Recognition, pp. 4401–4410 (2019)
19. Karras, T., Laine, S., Aittala, M., Hellsten, J., Lehtinen, J., Aila, T.: Analyzing and improving the image quality of StyleGAN. In: Proceedings of the IEEE/CVF Conference on Computer Vision and Pattern Recognition, pp. 8110–8119 (2020)
20. Kingma, D.P., Ba, J.: Adam: A method for stochastic optimization. In: Proceedings of the International Conference on Learning Representations, pp. 1–15 (2014)
21. Kingma, D.P., Welling, M.: Auto-encoding variational Bayes. In: Proceedings of the International Conference on Learning Representations, pp. 1–14 (2014)
22. Lee, H.Y., Tseng, H.Y., Huang, J.B., Singh, M., Yang, M.H.: Diverse image-to-image translation via disentangled representations. In: Proceedings of the European Conference on Computer Vision, pp. 35–51 (2018)

23. Lee, H.Y., Tseng, H.Y., Mao, Q., Huang, J.B., Lu, Y.D., Singh, M., Yang, M.H.: DRIT++: Diverse image-to-image translation via disentangled representations. International Journal of Computer Vision **128**(10), 2402–2417 (2020)
24. Li, X., Wang, X., He, X., Chen, L., Xiao, J., Chua, T.S.: Hierarchical fashion graph network for personalized outfit recommendation. In: Proceedings of the International ACM SIGIR Conference on Research and Development in Information Retrieval, pp. 159–168 (2020)
25. Liu, L., Zhang, H., Ji, Y., Wu, Q.J.: Toward AI fashion design: An Attribute-GAN model for clothing match. Neurocomputing **341** (2019)
26. Liu, L., Zhang, H., Xu, X., Zhang, Z., Yan, S.: Collocating clothes with generative adversarial networks cosupervised by categories and attributes: A multidiscriminator framework. IEEE Transactions on Neural Networks and Learning Systems **31**(9), 3540–3554 (2019)
27. Liu, L., Zhang, H., Zhou, D., Shi, J.: Toward fashion intelligence in the big data era: State-of-the-art and future prospects. IEEE Transactions on Consumer Electronics (2023)
28. Mao, Q., Tseng, H.Y., Lee, H.Y., Huang, J.B., Ma, S., Yang, M.H.: Continuous and diverse image-to-image translation via signed attribute vectors. International Journal of Computer Vision **130**(2), 517–549 (2022)
29. McAuley, J., Targett, C., Shi, Q., Van Den Hengel, A.: Image-based recommendations on styles and substitutes. In: Proceedings of the International ACM SIGIR Conference on Research and Development in Information Retrieval, pp. 43–52 (2015)
30. Paszke, A., Gross, S., Chintala, S., Chanan, G., Yang, E., DeVito, Z., Lin, Z., Desmaison, A., Antiga, L., Lerer, A.: Automatic differentiation in PyTorch. In: Proceedings of the Advances in Neural Information Processing Systems, pp. 1–4 (2017)
31. Richardson, E., Alaluf, Y., Patashnik, O., Nitzan, Y., Azar, Y., Shapiro, S., Cohen-Or, D.: Encoding in style: A StyleGAN encoder for image-to-image translation. In: Proceedings of the IEEE/CVF Conference on Computer Vision and Pattern Recognition, pp. 2287–2296 (2021)
32. Rombach, R., Blattmann, A., Lorenz, D., Esser, P., Ommer, B.: High-resolution image synthesis with latent diffusion models. In: Proceedings of the IEEE/CVF Conference on Computer Vision and Pattern Recognition, pp. 10684–10695 (2022)
33. Schroff, F., Kalenichenko, D., Philbin, J.: Facenet: A unified embedding for face recognition and clustering. In: Proceedings of the IEEE/CVF Conference on Computer Vision and Pattern Recognition, pp. 815–823 (2015)
34. Schuster, M., Paliwal, K.: Bidirectional recurrent neural networks. IEEE Transactions on Signal Processing **45** (1997)
35. Shen, Y., Yang, C., Tang, X., Zhou, B.: Interfacegan: Interpreting the disentangled face representation learned by gans. IEEE Transactions on Pattern Analysis and Machine Intelligence (2020)
36. Shi, J., Zhang, H., Zhou, D., Zhang, Z.: Toward intelligent interactive design: A generation framework based on cross-domain fashion elements. In: Proceedings of the ACM International Conference on Multimedia (2023)
37. Shu, X., Xu, B., Zhang, L., Tang, J.: Multi-granularity anchor-contrastive representation learning for semi-supervised skeleton-based action recognition. IEEE Transactions on Pattern Analysis and Machine Intelligence (2022)
38. Tewari, A., Elgharib, M., Bharaj, G., Bernard, F., Seidel, H.P., Pérez, P., Zollhofer, M., Theobalt, C.: Stylerig: Rigging stylegan for 3d control over portrait images. In: Proceedings of the IEEE/CVF Conference on Computer Vision and Pattern Recognition, pp. 6142–6151 (2020)
39. Tov, O., Alaluf, Y., Nitzan, Y., Patashnik, O., Cohen-Or, D.: Designing an encoder for StyleGAN image manipulation. ACM Transactions on Graphics **40**(4), 1–14 (2021)
40. Vasileva, M.I., Plummer, B.A., Dusad, K., Rajpal, S., Kumar, R., Forsyth, D.: Learning type-aware embeddings for fashion compatibility. In: Proceedings of the European Conference on Computer Vision, pp. 390–405 (2018)

41. Veit, A., Kovacs, B., Bell, S., McAuley, J., Bala, K., Belongie, S.: Learning visual clothing style with heterogeneous dyadic co-occurrences. In: Proceedings of the IEEE/CVF International Conference on Computer Vision, pp. 4642–4650 (2015)

42. Xia, W., Zhang, Y., Yang, Y., Xue, J.H., Zhou, B., Yang, M.H.: Gan inversion: A survey. IEEE Transactions on Pattern Analysis and Machine Intelligence **45**(3), 3121–3138 (2022)

43. Yan, R., Xie, L., Shu, X., Zhang, L., Tang, J.: Progressive instance-aware feature learning for compositional action recognition. IEEE Transactions on Pattern Analysis and Machine Intelligence (2023)

44. Yu, A., Grauman, K.: Fine-grained visual comparisons with local learning. In: Proceedings of the IEEE/CVF Conference on Computer Vision and Pattern Recognition, pp. 192–199 (2014)

45. Yu, C., Hu, Y., Chen, Y., Zeng, B.: Personalized fashion design. In: Proceedings of the IEEE/CVF International Conference on Computer Vision, pp. 9046–9055 (2019)

46. Zhang, H., Wang, X., Liu, L., Zhou, D., Zhang, Z.: Warpclothingout: A stepwise framework for clothes translation from the human body to tiled images. IEEE MultiMedia **27**(4), 58–68 (2020)

47. Zhang, R., Isola, P., Efros, A.A., Shechtman, E., Wang, O.: The unreasonable effectiveness of deep features as a perceptual metric. In: Proceedings of the IEEE/CVF Conference on Computer Vision and Pattern Recognition (2018)

48. Zhou, D., Zhang, H., Li, Q., Ma, J., Xu, X.: COutfitGAN: Learning to synthesize compatible outfits supervised by silhouette masks and fashion styles. IEEE Transactions on Multimedia pp. 1–15 (2023)

49. Zhou, D., Zhang, H., Ma, J., Fan, J., Zhang, Z.: Fcboost-net: A generative network for synthesizing multiple collocated outfits via fashion compatibility boosting. In: Proceedings of the ACM International Conference on Multimedia, pp. 1–9. ACM (2023)

50. Zhou, D., Zhang, H., Yang, K., Yan, H., Xu, X., Zhang, Z., Yan, S.: Learning to synthesize compatible fashion items using semantic alignment and collocation classification: An outfit generation framework. IEEE Transactions on Neural Networks and Learning Systems pp. 1–16 (2023)

51. Zhu, J., Shen, Y., Zhao, D., Zhou, B.: In-domain GAN inversion for real image editing. In: Proceedings of the European Conference on Computer Vision, pp. 592–608. Springer (2020)